CHRONOTYPES

Contributors

JOHN BENDER

CORNELIUS CASTORIADIS

DAVID WILLIAM COHEN

JOHANNES FABIAN

JACK GOODY

TAMARA K. HAREVEN

DOMINICK LACAPRA

THOMAS LUCKMANN

JONATHAN Z. SMITH

GAYATRI CHAKRAVORTY SPIVAK

BASTIAAN C. VAN FRAASSEN

DAVID E. WELLBERY

CHRONOTYPES

The Construction of Time

EDITED BY JOHN BENDER AND
DAVID E. WELLBERY

STANFORD UNIVERSITY PRESS
Stanford, California 1991

Stanford University Press, Stanford, California

CIP data appear at the end of the book
Printed in the United States of America

Contents

Acknowledgments

THIS VOLUME EMERGED from a conference given at Stanford University, February 13–15, 1988. The gathering, sponsored in large part by the Andrew W. Mellon Foundation, took place under the aegis of the Department of Comparative Literature, the Program in Modern Thought and Literature, and the Stanford Humanities Center. Stanford's Centennial Committee and the Dean of Graduate Studies assisted with substantial financing. The effort found inspiration in the receipt of a Mellon Foundation grant to foster "new combinations," one segment of which the School of Humanities and Sciences dedicated to a venture in interdisciplinary study.

The broad interdisciplinary scope of this volume, which touches numerous areas in the humanities, the social sciences, and the philosophy of science, took impetus from the efforts of a dedicated planning committee that included Russell Berman, Judith Brown, William Brown, Bliss Carnochan, Peter Galison, Van Harvey, Morton Sosna, Katherine Trumpener, Richard Terdiman, and Ian Watt. To their informed wisdom we owe the choice of a fine range of contributors. Barbara Mendelsohn's prize-winning poster captured our theme in compelling visual terms and provided inspiration for the jacket of this volume.

We especially wish to thank Bliss Carnochan and Morton Sosna, Director and Associate Director of the Stanford Humanities Center, for unfailingly generous advice and support from the most preliminary stages of planning, through the conference, to the execution of the volume itself. We are grateful also to the Humanities Center staff for smoothing out countless wrinkles, and especially to Dee Marquez, who typed the entire manuscript of this book. Marie Brazil saw to physical arrangements for the conference. Steven Brown contributed valuable substantive advice as well as necessary research and checking. Karen Rezendes, the unfailing mainstay of the Department of Comparative Literature, took care of financial and clerical details, conducted correspondence, and saw to the final manuscript as well as to many later aspects of book production. It is impossible to thank her enough.

Finally, we are happy to thank the contributors themselves for their timely responses to requests for finished copy and, then, for the seemingly inevitable changes to it.

J.B.
D.E.W.

Contributors

JOHN BENDER is Professor of English and Comparative Literature at Stanford University. He is author of *Spenser and Literary Pictorialism* and *Imagining the Penitentiary: Fiction and the Architecture of Mind in Eighteenth-Century England.* He recently coedited *The Ends of Rhetoric: History, Theory, Practice.*

DAVID E. WELLBERY is Professor of German at The Johns Hopkins University and Professor of German Studies and Comparative Literature at Stanford University. He is the author of *Lessing's 'Laocoon',* has edited *Positionen der Literaturwissenschaft,* and has coedited both *Reconstructing Individualism* and *The Ends of Rhetoric: History, Theory, Practice.*

CORNELIUS CASTORIADIS is Director of Studies at the Ecole des Hautes Etudes en Sciences Sociales in Paris. He is the author of *The Imaginary Institution of Society* and *Crossroads in the Labyrinth,* both published by MIT Press. His three-volume *Political and Social Writings* is in the process of appearing from the University of Minnesota Press.

DAVID WILLIAM COHEN is Director of the Program of African Studies, as well as Professor of History and Anthropology, at Northwestern University. He is the author of *Towards a Reconstructed Past* and, with E. S. Atieno Odhiambo, *Siaya: The Historical Anthropology of an African Landscape.* At present Cohen is completing, with Atieno Odhiambo, a book-length study of the 1986–87 struggle over the corpse of the Kenyan lawyer S. M. Otieno.

JOHANNES FABIAN is Professor of Cultural Anthropology at the University of Amsterdam. Previously he taught at Northwestern and Wesleyan universities and at the National University of Zaire. His publications include *Time and the Other, Language and Colonial Power, Power and Performance: Ethnographic Explorations Through Proverbial Wisdom and Theater in Shaba (Zaire),* and *History from Below.*

JACK GOODY is Fellow of St. John's College, Cambridge. He is the author of *The Domestication of the Savage Mind, The Logic of Writing and the Organisation of Society, The Interface Between the Written and the Oral,* and other works.

TAMARA K. HAREVEN is Unidel Professor of Family Studies and History at the University of Delaware and Member of the Center for Population Studies at Harvard University. She is also the editor of the *Journal of Family History.* Hareven has written numerous articles and books on the American family, the life course, and aging. One of her best-known books is *Family Time and Industrial Time.* She is currently writing a book entitled *Generations in Historical Time,* which focuses on issues of changes in the life course and in generational relations in American society.

DOMINICK LACAPRA is Goldwin Smith Professor of European Intellectual History and member of the field of Comparative Literature at Cornell University. His books include *'Madame Bovary' on Trial, Rethinking Intellectual History, History and Criticism, History, Politics, and the Novel,* and *Soundings in Critical Theory.*

THOMAS LUCKMANN is Professor in the Department of Sociology, University of Konstanz, Germany. His most important publications are *The Social Construction of Reality* (with Peter L. Berger), *The Invisible Religion, The Structures of the Life-World I* (with Alfred Schütz), *The Sociology of Language, The Structures of the Life-World II* (with Alfred Schütz), and *Life-World and Social Realities.*

JONATHAN Z. SMITH is the Robert O. Anderson Distinguished Service Professor of the Humanities, coordinator of the Religion and the Humanities Program, and a member of the Committee on the History of Culture at the University of Chicago. His most recent works include *To Take Place: Toward Theory in Ritual* and *Drudgery Divine: On the Comparison of Early Christianities and the Religions of Late Antiquity.*

GAYATRI CHAKRAVORTY SPIVAK is Andrew W. Mellon Professor of English at the University of Pittsburgh. She is the translator of Jacques Derrida, *Of Grammatology,* and the author of *In Other Worlds.* Her forthcoming book is *Master Discourse, Native Informant: Deconstruction in the Service of Reading.*

BASTIAAN C. VAN FRAASSEN is Professor of Philosophy at Princeton University. His books include *The Scientific Image, Laws and Symmetry,* and *Quantum Mechanics: An Empiricist View.*

CHRONOTYPES

Introduction

JOHN BENDER AND DAVID E. WELLBERY

TIME BELONGS TO a handful of categories (like form, symbol, cause) that prompt universal concern. Time touches every dimension of our being, every object of our attention—including our attention itself. It permeates simple everyday experience no less than the most abstrusely theoretical speculation. Time therefore can belong to no single field of study. In contrast to notions such as "tragedy," "quark," or "market," which have distinct disciplinary locales, time confronts every specialty. Its relevance extends from the strictest computations of mathematical science to the most speculative conjectures of interpretive theory. It is a genuinely transdisciplinary category.

Of course, this universalist view of time is not itself universal but rather the product of the modern age, of an age that conceives itself, in temporal terms, as the "new" time. In his book *Futures Past: On the Semantics of Historical Time,* Reinhart Koselleck argues that the late eighteenth and early nineteenth centuries constitute a historical threshold during which emerged a conception of unified and all-pervasive change as occurring in and through time.[1] This *temporalization* of experience—this notion of time as the framework within which life forms are embedded and carry on their existence—is the defining quality of the modern world. Only where such a framework is presupposed can something like history as the collective unity of all individual event sequences be conceived. Time loses its character as a locational marker and becomes the productive medium that generates, at an accelerating rate, innovative experiential configurations. Evidence outside the historical theories studied by Koselleck corroborates his view that time is the *a priori* of the modern world. One need only think of the philosophical centrality of time (subjectively conceived) from Kant to Heidegger; of the reconceptualization of geological history, whose temporal parameters have been studied by Stephen J. Gould; of the introduction into physics, through the second law of thermodynamics, of a concept of irreversible time.[2] Moreover, modern social life rests on increasingly exact mechanisms of temporal coordination, and our technical media enable, for the first time in history, the recording, storage, and reproduction of kinetic phenomena.

We note these matters here in order to highlight what we take to be the overarching intellectual context of the present volume. It is often claimed that our contemporaneity should be characterized as postmodern, a predicate that becomes more slippery with proliferating use. If, however, the term denotes an exponential intensification of modernity—a process whereby the underlying structures of modern thought and experience become self-reflexive—then it signifies something useful and does illuminate the present endeavor. The essays collected here may be said to participate in a postmodern style of inquiry because they evince, from various disciplinary perspectives, a recognition of the problematic and complex standing of the temporal categories that lent those disciplines their modern shapes. In this sense, these essays are symptomatic of a general turbulence of thought now manifesting itself in many areas of intellectual endeavor. The comparatively recent emergence of time as a central theme of research is no accident, but rather a consequential extension of the postmodern turn: our present does not leave modernity behind, but rather aggravates its difficulties, intensifies its concerns.

The interdisciplinary character of the volume reflects another aspect of the postmodern intellectual climate. Because the theme of time moves through varied disciplines, it offers marked opportunities for collective discussion, throws points of communication into relief, and displaces the neatly parceled topography that academia ordinarily exhibits. To be sure, the interdisciplinary discussion carried on here does not include voices from the natural sciences. It leaves unaddressed such questions as the temporal paradoxes of modern physics or the intricacies of geological time and biological clocks. Rather, the main concern of these essays lies with questions of time as treated in the social sciences and humanities, that is, in fields Europeans properly term the "human sciences." The volume does begin, however, with two essays that in different ways negotiate passages between scientific and humanistic perspectives. Moreover, all of the authors acknowledge the inseparability of human chronotypes from the temporal patterns of the physical, the chemical, and the biological.

One simplification of this book would be to designate narrative as its central subject matter. Indeed, the notion of "chronotypes" is a variation on a concept in narrative theory. Mikhail Bakhtin borrowed the suggestive term "chronotope" from Einstein's physics in order to designate the fusion of temporal/spatial structures and to define characteristic time/space formations in specific narrative genres such as the romance, the idyll, the folktale, the picaresque novel, and so forth.[3] We have modified Bakhtin's term to make explicit its typological function and to focus upon the concept of temporality. Bakhtin's term is suggestive because it points to the diversity of prototypical cultural forms within which time assumes significance. Certainly no consideration of narrative could avoid the question of time, and quite probably the converse also holds true. This, at least, is the implication of the essays gathered here. The authors repeatedly depict time as narrative construction and themselves adopt distinctive narrational attitudes in presenting their own evidence and argumentation. The standing of narrative remains continuously in question.

If, as contemporary observers like Jean-François Lyotard and Niklas Luhmann have suggested, the turbulence of thought now evident in many fields of investigation points symptomatically to an epistemic or paradigmatic shift in the basic categories through which human life forms construct metropolitan reality, we could well expect naturalized conceptions of time to be overturned.[4] And when the routinely accepted shapes of time that form the backdrop of everyday practice fall into the line of analytic vision, narrative also demands attention. Indeed, the concurrent reconfiguration of temporal categories and narrative procedures appears to operate with a virtually systemic logic, at least in the history of Western culture.

But the drive to comprehend temporal construction as a function of narrative formation may now be assuming a historically specific urgency—and therefore a place of significance in this volume—because the Newtonian coordination of time and space as abstract constants within a measurable framework grows ever more fragile as the theoretical truths of relativity converge during the late twentieth century with the subjective pressures of instantaneous digitalized communication over vast spatial, temporal, and cultural spans. Donald J. Wilcox has shown how the older world of multifarious narratively and rhetorically structured chronologies was displaced by the conception of absolute time as elaborated by Descartes and Newton in the seventeenth century and symbolized by the spread of the B.C./A.D. dating system.[5] Operating under the hegemony of absolute time, narrative was increasingly reduced to heuristic status and, whether in the medium of the realist novel as it emerged in the eighteenth century or that of nineteenth-century historicism, was itself subordinated to a grid of abstract coordinates. Now, with the destabilization of the Newtonian grid, narrative reasserts itself as problematic and we confront a future inhabited by multiple times.

The coinage "chronotypes" implies a number of presuppositions that, together, set an agenda for the essays that follow. Chronotypes are models or patterns through which time assumes practical or conceptual significance. Time is not given but (as our subtitle indicates) fabricated in an ongoing process. Chronotypes are themselves temporal and plural, constantly being made and remade at multiple individual, social, and cultural levels. They interact with one another, sometimes cooperatively, sometimes conflictually. They change over time and therefore have a history or histories, the construal of which itself is an act of temporal construction. These things said, however, one caution also should be kept in mind. Chronotypes are not produced *ex nihilo;* they are improvised from an already existing repertoire of cultural forms and natural phenomena.

We can more fully delineate the thematic horizon that the general concept of chronotypes establishes for this volume by noticing a range of questions that run through all of the essays.

(1) How does the process of temporal construction occur? What are its instruments and procedures? What are its enabling technologies? In what ways does the fabrication of chronotypes build on, expand, and modify culturally available shapes of time?
(2) What functions do chronotypes serve? How do they contribute to the formation of social, cultural, and individual identity? How do they organize behavior? In what ways do they constrict or expand the field of experience?
(3) What is involved in knowing temporal construction? What idealizations and projections—whether conscious or unconscious—take place in the analysis of chronotypes? How does the observer's own position in time impinge on the process of description?
(4) Should narrative, as we already have implied, be treated as a privileged or notably efficacious chronotype? Is narrative a universal device for the organization of time or one specific to certain historical and cultural contexts? How is narrative linked to non-narrative chronotypes?
(5) What is the relation between temporal construction and empowerment? Are chronotypes involved in processes of domination?

Openings to these and other questions occur in the individual contributions. Taken together, these essays mark the emergence of a new transdisciplinary analytic to which the idea of temporal construction is central.

"Thinking Time," the first part of the volume, brings together contributions by two philosophers who address the question What is time? Their pursuit of an answer exemplifies the possibilities for interdisciplinary communication that the theme of time offers. The authors represent divergent styles of philosophical inquiry, Bastiaan C. van Fraassen the analytic direction in Anglo-American philosophy of science and Cornelius Castoriadis the Continental tradition of philosophical critique. Their essays provide a model for dialogue between philosophical schools that have remained insulated from one another. But more importantly, both authors situate their contributions at a point between the natural and the human sciences, between physics and fiction, the objective and subjective worlds. Where questions of time are at issue, these entrenched divisions begin to dissolve.

Van Fraassen performs a cross-reading by elucidating problems of fiction in terms of problems in physics, and vice versa. The crucial matter that allows for this juxtaposition is the question of temporal determination. How do we fix or define an event's location in time? Of course, as long as we hold with Newton that time is real, an actually existing thing, the issue could hardly arise. This idea is deeply embedded in common sense and common parlance. The temporal position of an event is simply the unique slice of time it occupies. Since Leibniz, however, this notion of time as an existing entity has suffered considerable erosion and an alternative conception has emerged, which van Fraassen calls the relational theory of time (or, in the context of modern physics, space-time). On this view, temporal order is a function of the relations that establish the compatibility or incompatibility among events.

Van Fraassen's discussion of the relational theory of time shows that, within its framework, severe problems of determination arise. These problems do not render the theory void, however, for it is just a fact of the matter that, in our constructions of the real world, indeterminacies cannot be entirely eliminated. And certainly, with respect to fictional worlds, the temporal localization of events likewise remains nebulous. Van Fraassen makes this point through reference to several examples, most notably perhaps that of Proust's *A la recherche du temps perdu,* which has long been a privileged text for literary critics concerned with questions of time. Thus, reality and fiction would seem to share, in principle, a temporal indeterminacy that subverts our ability rigorously to distinguish the two domains. Reality, like fiction, is a text in which assignments of temporal position cannot be finally decided. Van Fraassen's paper thus develops a perspective from which the empiricism of the philosophy of science and the textualism of current literary criticism could enter into productive discussion.

Castoriadis begins with the inherited distinction between subjective and objective time, time as constructed by beings who exist for themselves and time as a dimension of the natural world. He works to bridge the gap between these two kinds of time, or rather to develop a notion of time as such that would allow us to recognize both forms as related to each other.

In a first step, Castoriadis examines Aristotle's and Augustine's concepts of time, elaborating the basic features of the objectivist paradigm established by the former and of the subjectivist model developed by the latter. Neither conception can stand on its own. Both Aristotle and Augustine must introduce into their accounts of time elements that belong to the competing theory. The strict distinction between subject and object that has so powerfully shaped our theoretical traditions quite simply doesn't hold. But Castoriadis's major criticism of that tradition bears on another matter: the persistent neglect of the sociohistorical dimension of reality. And here, in particular, the untenability of the division between subjective and objective worlds appears. For what would the sociohistorical world be if it did not lean on and draw its resources from a natural world that preexists it? And what would the natural world be if it were not shaped and altered by social and historical forces, if it were not known through instruments history produces? Castoriadis's discussion of the social world reveals the interlacing of subjective and objective domains especially with regard to the construction of time. In this respect his paper sets a frame for several that follow. In his view, time is a social dimension shaped and reshaped in a network of chronotypes.

And yet, what is time if neither subjective nor objective? What form must time assume if it leashes together the natural and social worlds? Castoriadis's answer to this question is twofold. Time is the form of diversity, and as such it manifests itself in two ways. First, it organizes diversity as difference, as measurable distinctions governed by a single law. Time conceived in this way exhibits uniformity; it is quantifiable and regular. Such would seem to be the time described in physics, which van Fraassen's paper discusses. For Castoriadis, however, time must be conceived also in terms of a type of relation other than difference, a relation of diversity that knows no common measure. He calls this relation "otherness," a radical heterogeneity. Whenever new forms emerge, be it in nature, society, or individual experience, otherness happens. In other words, Castoriadis argues that time is *alter*-ation, the becoming of alterity, the creation and destruction of forms. At this most fundamental level, time resists homogenization and calls for the construction of chronotypes capable of grasping the emergence of the other and the new.

In Part II, "Temporal Frames of Inquiry," Jonathan Z. Smith and Jack Goody center attention on the distorting power of paradigms that govern inquiry. They speak from the distinct yet interrelated disciplinary contexts of two broadly influential fields, the history of religion and cultural anthropology. One deals with Middle Eastern antiquity, the other with tribal cultures of this century. But both are concerned with the problem of temporal idealization, with the usefulness and the mendacity of our investment in neat frames and either/or contrasts. Both confront the power of ideology to shape our views of the past, of the "other," and therefore of ourselves.

Smith focuses on the temporality implicit in modern paradigms of inquiry. He debunks the great divide between the worlds of pagan cyclical time and Hebreo-Christian historical and linear time. This duality presents a before-and-after relation pervasive in religious studies. Smith argues that this duality is an invention of the mid–nineteenth century rather than characteristic of religion itself. He pays particular attention to the articulation of this formula (one might even say "mythology") among a group of German scholars known as the Pan-Babylonian School, but also in Durkheim and Eliade. He finds that while several influential temporal schemata purport to map the phenomenology of religion in distinctive fashion, they all place great cultural entities in before-and-after relationships. In this sense they replicate the basic chronotype of Christianity at its foundation and, most especially, at its reinauguration in Protestant theology.

Smith sees this replication as fundamentally corrosive to historical inquiry. At least implicitly, he throws into question the structuralist contention that a fundamental binarism governs human experience, culture, and thought. If the binary structure that has been taken as so fundamental to our understanding of religious history is in fact an epiphenomenal product of Protestant values as refracted through nineteenth-century historicism, then structuralism itself might be counted as a tributary to this discourse. Smith could be seen as historicizing binarism itself (at least as it governs modern religious studies). He might also be seen as questioning whether our construction of historical time in terms of epochs may in fact be a paradigm colored by European religious formation. In this regard, his contribution echoes the critique of epochal history developed in Gayatri Chakravorty Spivak's paper.

An abiding assumption of cultural anthropology has been that oral cultures are narratively dominated. Goody questions this traditional view and stresses instead the discontinuities imposed by real-time exigencies in oral culture, where boredom and competition may confine extended narratives to specialized, often ritual, contexts. Interpersonal necessities come into play, such as the individually felt need of group members to speak off the subject or to supply irrelevant detail. Similarly, the collective requirement to speak in socially prescribed sequence or tonality often takes priority over narrative organization. The time available to tell a story emerges, as in Johannes Fabian's essay later in the volume, as a commodity subject to competition and power relations. Goody shows that several features usually associated with "narrative" in fact are functions of writing, not of orality, and that the precise temporal framing and cross-referentiality ordinarily associated with narrative are uncharacteristic of story telling in oral cultures. Writing may, in fact, lock contemporary anthropological inquiry into the acceptance of a culture's master narrative or dominant story, whereas traditional oral cultures can keep many stories alive in variously continuous and discontinuous forms.

Goody probes the effects of understanding difference (in this case between orality and literacy) in terms of a single contrastive temporal grid. He suggests that our styles of inquiry, aimed as they are at written production, reduce the oral and keep down the noise it characteristically tolerates by selecting from it those narratives that (1) are capable of linearization, (2) are chosen by those in power, and (3) meet the modern researcher's need to find material that can be structured into the narrative form of the modern research report. Like Fabian, he questions the narrative forms that have dominated modern research and shows how the reductive nature of our economical, linear packaging virtually forces us to dismiss the clamor of oral exchange as insignificant. What we may have found in oral cultures is our own narrative boundedness—our own chronotype.

The essays in Part III, "Time and the Politics of Criticism," work through issues bearing on literary and cultural criticism. Insofar as it attends to historical practices and artifacts, criticism is tied up with questions of time. It must construct a relation between the "then" and the "now"; it must construe patterns of historical change; it must reflect on the inner temporality of the texts it attends to. As the essays by Gayatri Chakravorty Spivak and Dominick LaCapra show, these operations include significant ideological and political components.

Spivak develops her argument with reference to the most influential modern construction of historical time, that of Hegel. In particular, she analyzes his charting of the history of art forms in the *Lectures on Fine Arts.* In line with his dictum that time is the existence of the spirit, Hegel developed in his aesthetic theory a thoroughgoing temporalization of art, and there can be little doubt that the Hegelian chronotype maintains a hold on our conceptions of literary and art history. In Spivak's view, this chronotype functions by subsuming the diversity and fragmentation of temporal experience—what she calls "timing"—to the monotone unfolding of the one universal law (Time). She discusses Hegel's comments on the *Srimadbhagavadgita,* illustrating his reduction of the complexity and historical tensions of Indian culture to a single moment within his schematization of the spirit's itinerary toward self-knowledge. Spivak's remarks here may be compared to those of Castoriadis, who refers to Hegel in passing as a thinker of the homogeneous time of difference as opposed to the creative time of otherness.

The proximity of Spivak and Castoriadis becomes all the more apparent when, in the second phase of her paper, Spivak develops a reading of the *Gita* itself as a temporal construction that operates very much like the Hegelian chronotype. To put the matter simply (a simplification which the reader of Spivak's essay will correct), the grand lesson of the god Krishna to the prince Arjuna absorbs the fragmented temporality of phenomenal experience into an omnitemporality governed by the law the god represents. This chronotypic transformation then serves to legitimate the order of the castes, which are named at the end of the episode. In short, the poem projects a temporal construction that itself sustains an emergent form of state domination; it is, as Castoriadis would say, a product of the sociohistorical imaginary. This point holds, albeit in a different way, for Hegel as well. The Hegelian chronotype enables the consolidation of the modern nation state.

Another way of phrasing this is to say that Hegel's theory of history, and of the history of art, serves colonialist interests. As Spivak shows, however, the same Hegelian construction which allocated to India the role of the exotic and static Other of European progress was itself adapted by Indian nationalists in the assertion of their own authentic identity. A sort of mirror game of self and other is set into motion here, obeying dynamics that Johannes Fabian also describes from the standpoint of anthropology. Spivak endeavors to step outside this specular exchange by deconstructing the opposition between colonizer and colonized. The issues—including those of temporal construction—are much more intricately intertwined than this simple dichotomy suggests. Spivak's discussion of this problem should be compared with the argument of David William Cohen in his essay on La Fontaine and Wamimbi.

LaCapra's essay focuses on two recent, highly influential accounts of Romanticism. Interestingly, his first example, M. H. Abrams's study *Natural Supernaturalism,* endorses the Hegelian chronotype discussed by Spivak. At the basis of Romanticism Abrams finds the Hegelian parsing of historical change as a movement from original unity through disunity and alienation to a recovered unity characterized by self-knowledge. The rhetorical figuration that ratifies this chronotype is the symbol, in which the unity and totality posited as both origin and end achieve phenomenal realization. The major portion of LaCapra's discussion, however, is devoted to Paul de Man's essay "The Rhetoric of Temporality," the title of which LaCapra's own inverts. De Man's theory characterizes the Romantic achievement as the insight into temporal discontinuity and therefore privileges the strategies of allegory and irony that, with different but related accentuations, enact this insight rhetorically. From this perspective, the symbol, with its promise of unity, appears as a mystification, as does a construction of literary history that integrates the Romantic rupture within a continuous narrative. It would seem, then, that de Man offers the possibility of constructing historical time—and in particular the time of literary history—in a fashion that avoids the Hegelian leveling of otherness.

LaCapra recognizes this potential in de Man's essay, but he by no means subscribes entirely to the de Manian position. On the contrary, his comparison of de Man and Abrams reveals that their theories can be viewed as mirror reflections of each other, that Abrams's exclusive insistence on organic continuity is countered in de Man's equally exclusive valorization of disjunction and absence. This specular standoff has its counterpart in the binarism that internally organizes each critic's argumentation, and LaCapra's analysis of these conceptual and figural dualities reveals the ideological investments that support, and are supported by, Abrams's and de Man's constructions of time.

LaCapra's project, however, is not solely critical. His ultimate aim—one he shares with several contributors to the volume—is to work through the phase of negative criticism in order to disclose possibilities of temporal articulation that are not reducible to binary oppositions. A central category he develops in this regard is displacement: the reemergence in an altered or distorted form of a submerged or repressed practice. As LaCapra points out, de Man's own essay can be read in this sense as a displacement of Protestant asceticism, a contention that communicates with Jonathan Z. Smith's argument regarding nineteenth-century theories of religion. Or to cite a second example, Baudelaire's essay on laughter, a text de Man interprets, displaces carnivalesque and hermeticist elements that de Man's formalism compels him to ignore. LaCapra's notion of displacement—and it is only one example of those he explores—adumbrates a historical chronotype that escapes the opposition of continuity and disjunction.

In "The Temporal Order of Social Life," the fourth part of the volume, Thomas Luckmann and Tamara K. Hareven show that from the perspectives of sociology and social history we ought to speak not of time but of multiple timings or of clocks running simultaneously. Both stress not only individual temporal awareness but the many ways in which time is asynchronously produced by physical circumstance, historical events, social placement, and the expectations of family or other communal units.

Luckmann is interested in the multiplicity, relativity, and intersubjectivity that characterize individual time. The time of reflection or daydream differs, for example, from the awareness of time in everyday life, which is structured by externally imposed conditions and by the interventions of others. The confluence of shared expectation and bodily presence may blur the time of one individual into that of another through the correlation of streams of consciousness. Institutions and laws also work to situate individuals within various collectively derived temporal orders. One thinks of Castoriadis's account of the social phenomenology of time when Luckmann says that everyday life is ordered within "socialized intersubjective time." But he also shows the relativity of social time itself. It consists not of pre-made or self-made categories but, simultaneously, of uncodified practice (experienced by individuals as normal or commonsensical) and of abstracted knowledge about objectified time and temporal categories of interaction in society. Biographical schemes, on Luckmann's account, form narrative bridges that work to integrate widely varying individual and collective time schemes. These embrace any number of short-term or long-term—fast or slow—temporal sequences, including, of course, the comparatively objective readings of mechanical and biological clocks.

Hareven's study is broadly compatible with these generalizations. She discusses ways in which the "life course" as socially defined and individually experienced by specific cohorts of factory workers and their children in New England and Japan allows us to understand generational changes in the synchronization of individual time, family time, and historical time. In Hareven's view the family serves as "an important mediator between individuals and the grander social processes." She analyzes the perceived timeliness—in different generations and cultures—of such transitional moments in the life course as school completion, first employment, marriage, setting up a separate household, and childbearing. She considers how individuals view the timing of these transitions as well as how their perceptions and needs are synchronized with those of the collectivity and transmitted from generation to generation. Different generations define the "turning points" in their own life courses in strikingly different ways. "Early" or "delayed" timing of transitions because of historical events like war or economic depression may affect the overall life course. For early-twentieth-century immigrants to a New England textile mill, the significant turning points most often were marked by historical events, business cycles, and moments in the work career. For their children, the turning points more often were family events or personal crises. Hareven's findings may point to an issue raised by various contemporary social theorists: the characteristically large distances in time/space between elements constituting the huge social formations typical of the modern industrialized West put a premium on coordination of individual, family, and societal time while paradoxically rendering coordination extremely difficult.

Luckmann and Hareven, like Fabian and Cohen, consider narrative schemes as conventions of thought and social organization that not only account for time past but structure time present and time future. Luckmann joins Fabian in showing how narrative relativizes not only personal and societal time but past, present, and future. States like pastness and futurity become fluid and indeterminate in ways akin to those described by van Fraassen.

The essays in the final part of the volume, "Time, Narrative, and Cultural Contact," touch on issues developed in previous contributions. Both Johannes Fabian and David William Cohen are concerned with the operations involved in temporal construction and with their ideological implications; both address the function of time in the dialectic of self and other; both investigate techniques of narrative and ambiguities of temporal location. The overall context within which these questions are addressed is the contact of cultures as it pertains to the disciplines of anthropology and history.

Fabian employs two anecdotes from personal experience in order to illustrate mechanisms of temporal construction that play a significant role in the production of anthropological knowledge. The first of these ("Of Dogs Alive") recalls van Fraassen's observations on the relational theory of time by demonstrating that even our "present"—our subjective standpoint at a particular place and time—is fabricated through the establishment of a relational network. The indexical expressions ("we," "now," "here") with which we refer to this standpoint, and which seem merely to point out a self-evident state of affairs, in fact selectively segment the world. By removing the veil of naturalness that covers our constitution of the present, Fabian opens this most immediate form of temporal construction to the possibility of ideological critique. Other presents, other we's, heres, and nows, are possible: where we choose to live and work is an essentially political question.

Fabian's second anecdote involves the experience of an omen, a sudden and startling event that seems to portend ill or fortunate consequences. In Fabian's analysis, the apprehension of an event as ominous illustrates the more general process of rendering the world meaningful through narrativization. An omen is constituted as such when it is seen, even in the moment we experience it, as a "past" that will find confirmation in a "future." Otherness is domesticated and at the same time distanced from us, through its integration into a narrative chronotype. Here too Fabian discloses the ideological implications of the temporalization he investigates. The taming of strangeness that occurs in the identification of the omen suggests a more general process through which traumatic, incomprehensible, or conflictual encounters can be narratively smoothed over and repressed. Something akin to the phenomenon of "deferred action" (*Nachträglichkeit*) described by Freud—also alluded to by La-Capra—seems to be operative here.

Fabian's overall concern, as mentioned above, bears on the question of how the temporal constructions he designates as "time-space fusion" and "divinatory narrativization" are involved in the work of anthropologists. In his view, anthropologists often construe their present position so as to prevent access to contemporaneity by the cultures they study. And they often develop narrative schemata—theories of cultural evolution, for example—that homogenize otherness and efface the violence and oppression that mark the history of cultural contact. But the critique of anthropology Fabian envisages is conceived as practical rather than conceptual. At the end of his paper he calls for an examination of anthropological writing that would analyze devices, narrative and otherwise, that enable the anthropologist's work. David William Cohen's contribution begins to realize this program.

Indeed, Cohen opens his essay with a series of references to an earlier work by Fabian. In particular, he draws on Fabian's insight that ethnographic writing enacts a conquest of time. Cohen's approach to this general issue, however, involves a significant redrawing of the frame of inquiry. By analyzing both the text of a professional ethnographer, Jean La Fontaine's *The Gisu of Uganda* (1959), and the text by the Mugisu schoolteacher G. W. Wamimbi, *Modern Mood in Masaaba* (1970), Cohen delineates a complex dialogical interchange between the "scientific" and "native" points of view. It turns out that the conquest of time does not run in a single direction and that anthropological knowledge can serve, in unpredictable ways, to empower those very individuals the ethnographer describes.

Not only do La Fontaine's ethnography and Wamimbi's historical oration share the same field of reference, but the latter also draws on the former, expanding and transforming its perspective in myriad ways. Cohen conducts a detailed comparison of the two texts, focusing especially on passages that describe ritual circumcision. His analysis shows how the ethnographic study, with its dichotomy of time present and time past, is rehistoricized both explicitly and implicitly in the schoolteacher's account. Through a series of temporal alterations, Wamimbi reconquers time, constructing a narrative chronotype that allows for the depiction of contingency, variability, agency, and affectivity. Cohen argues that the dialogical process exemplified in these two texts illustrates the complexities involved in the interaction of oral and literate cultures, a theme central to Goody's paper as well. And his final remarks suggest an interpretation of the circumcision ritual discussed in La Fontaine's and Wamimbi's texts as itself a process of temporalization.

At the outset of this introduction we emphasized that the theme of time opens possibilities for interdisciplinary dialogue. This is amply demonstrated by the papers collected here. Time is once again at the center of the thematic field in significant areas of research. This concern for questions of temporal constitution bespeaks a general shift in research strategies. Structuralist trends, which so decisively influenced the human sciences beginning in the late 1960s, usually ignored questions of temporality in favor of synchronic analysis. Spatial metaphors—drawn from geology, geography, and topology, for example—organized thought about discourse and social formation. The predominance of spatial models in structuralist thinking called attention to issues that had been repressed in the monolithic temporal framework posited by Hegelianism and its historicist avatars. Similarly, structuralism repudiated the predominant stress on the inner flow of conscious temporal experience that was common to such movements as Bergsonism and phenomenology.

Perhaps one could say that the structuralist intervention broke up the continuity of time, but at the cost of freezing it or banishing it altogether from discussion and investigation.

The thematization of time in contemporary research draws to some degree on the insights of historicism and phenomenology, but is distinguished from these theoretical antecedents by the emphasis it places on plurality and complexity. Time as conceived by the contributors to this volume is not a single medium of consciousness or a unified movement in history. It is intrinsically manifold. Numerous chronotypes intertwine to make up the fabric of time. The social and cultural processes of temporal construction rely on and also reelaborate antecedent rhythms and articulations. These multiple times can become objects of contention because individuals experience them differently and because they bear ideological implications. Time asserts itself in contemporary inquiry less as a given than as a range of problems, the solutions to which are constantly open to renegotiation. For this reason, it would be inappropriate to conceive this collective endeavor as signaling the emergence of a new temporal paradigm across disciplines. The postmodern reflective turn tends rather toward the diversification of models and metaphors, toward a multiplication of times. Postmodern research and interdisciplinary discussion are not governed by a totalizing conception—unless it be the concept of postmodernity itself—but rather make do and invent in the postlapsarian awareness of paradigms lost. The insistent plural of our coinage "chronotypes" attempts to capture this aspect of contemporary research on time.

Among the first efforts in the Western tradition to think the nature of time were Zeno's famous paradoxes of the arrow and of Achilles and the tortoise. Today, most philosophers would agree that these paradoxes have been solved, their logical conundrums eliminated. Nevertheless, as images or narratives they continue to exert a fascination on us, commanding our attention as if they were allegories of our own situation. By pulling the ground of certainty from beneath us, by tangling us in their enigmas, they communicate something essential: the enigmatic character of time itself. Like Zeno's Achilles, pursuing the lethargic but deliberate movement of the tortoise, our thoughts about human time never seem to reach the bottom of the matter. Quite possibly, there is no bottom, no first or final answer to our insistent questions. The present volume suggests that this is not necessarily a bad situation, that time is intrinsically plural, and that this plurality is the very condition of our life and thought.

PART I

Thinking Time

Time in Physical and Narrative Structure

BASTIAAN C. VAN FRAASSEN

WHEN THE READER turns to a text, he conceives of the narrated events as ordered in time. When the natural philosopher turns to the world, he also conceives of its events as ordered in time—or lately, in space-time. But each has the task of constituting this order on the basis of clues present in what is to be ordered. Interrogating the parallels to be found in their problems and methods, I shall argue that in both cases the definiteness of the relation between the order and what is ordered resides mainly in how the matter is to be conceived, and is underdetermined by the facts.

Constructing Order in Narrative Time

Everyone expects an analytic philosopher to be analytic. So I shall start by taking as example the most analytic story I know. It is Dino Buzzati's "The Seven Messengers." Here are a few excerpts:

(1) Having set out to explore my father's kingdom, I go on day after day, drawing away from my city, and the news that reaches me becomes increasingly more infrequent.

I began the journey when I was little more than thirty years old, and more than eight years have passed, exactly eight years, six months, and fifteen days of uninterrupted travel. I believed, at my departure, that I would have easily reached the borders of the kingdom in a few weeks, but I have continued to encounter always new people and regions.

(2) Although carefree—much more than I am now!—I was preoccupied with the possibility of communicating with my family during the journey, and from the knights of my guard I selected the seven best to serve as my messengers.

Ignorant of my real situation, I supposed having seven of them was an utter extravagance. As time passed I perceived that on the contrary they were ridiculously few; and yet none of them has ever fallen ill, or run into brigands, or ridden his horse to death.

(3) To distinguish them easily, I gave them names with alphabetical initials: Alessandro, Bartolomeo, Caio, Domenico, Ettore, Federico, and Gregorio.

Unaccustomed to being away from my home, I dispatched the first, Alessandro, as early as the second night of the journey, when we had covered eighty leagues. The night after, to assure the continuity of the communications, I sent the second one, then the third, then the fourth, consecutively, until the eighth night of the journey, on which Gregorio departed. The first had not yet returned.

He arrived on the tenth night while we were pitching camp in an uninhabited valley. I learned from Alessandro that his speed had been inferior to my expectations: I had thought that proceeding alone, he could cover a distance twice ours in the same time; instead he made only one and a half. In one day, while we advanced forty leagues, he devoured sixty. . . .

I very quickly noted that it was sufficient to multiply by five the days elapsed thus far to know when the messenger would catch up with us.

(4) But eight and a half years have passed. Tonight I was having supper alone in my tent when Domenico entered, still able to smile though overcome with fatigue. For almost seven years I had not seen him. Throughout this very long period, he had done nothing but hurry, across grasslands, woods, and deserts. . . .

He will leave again for the last time. In my diary I have calculated that if all goes well, if I continue my journey as I have done till now and he continues his, I will again see Domenico only when thirty-four years have passed. I will then be seventy-two years old. Yet I begin to feel weary, and it is probable that death will seize me before that time. So I shall never see him again.

(5) You are the last link with them, Domenico. The fifth messenger, Ettore, who will reach me, God willing, in a year and eight months, will not be able to leave again because he would never have enough time to return. After you, silence, O Domenico, unless I finally find the longed-for boundaries. But the more I proceed, the more I become convinced that the frontier does not exist.[1]

The narrator of "The Seven Messengers" structures the time sequence with an algebraic formula. Perhaps because I stem from the culture of Vermeer and Mondrian, I have translated this formula into geometry (see Figure 1).

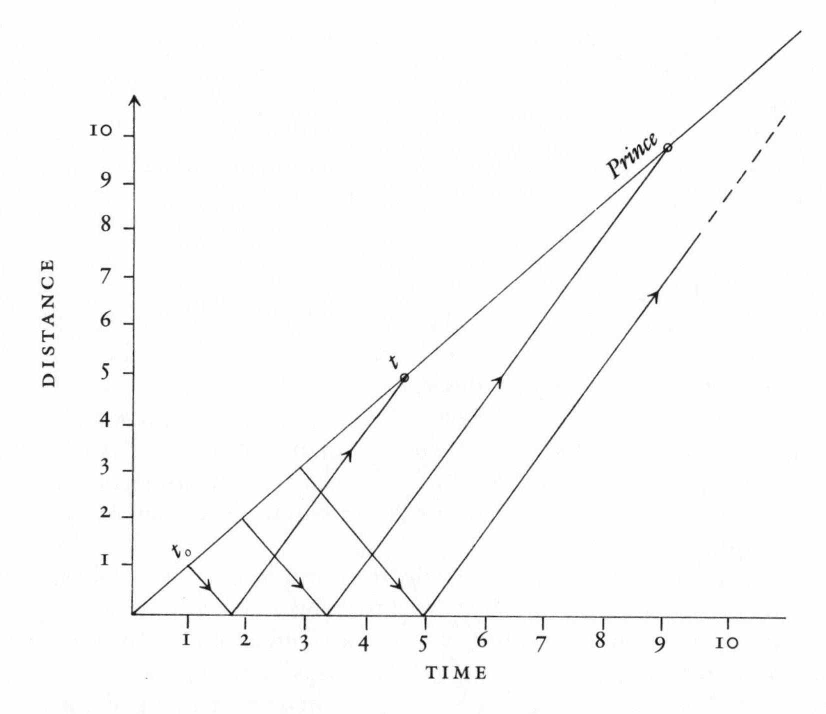

Unit of distance: one *princeling* = one Prince's day's journey
Messengers' speed: 1.5 times the Prince's speed
Prince-line: slope 1:1 (i.e., speed 1 princeling per day)
Messenger-lines: slope 1.5:1 (i.e., 1.5 princelings per day)

Distance traveled by a messenger must equal both:
(a) return journey to capital (t_o units) *plus* journey to catch up with the Prince (additional t units)
(b) time messenger spends away from Prince, multiplied by messenger's speed.
Thus, $(t + t_o) = 1.5(t - t_o)$, and therefore, $t = 5t_o$.

Messenger return days (Domenico is indicated by [ ])

2	3	4	[5]	6	7	8	
10	15	20	[25]	30	35	40	{1–6 wks}
50	75	100	[125]	150	175	200	{1–7 mos}
250	375	500	[625]	750	825	1,000	{8 mos–3 yrs}
1,250	1,875	2,500	[3,125]	3,700	4,125	5,000	{3–12 yrs}
. . .			[15,625]	. . . .			42.9 yrs

Fig. 1. The journeys in Dino Buzzati's "The Seven Messengers."

The Prince narrates this story in his thirty-eighth year. The last messenger he could ever hope to see return will leave tomorrow; the next to arrive, Ettore, will make no further trips to the capital. In only eight and a half years, the Prince has placed himself at an unbridgeable distance from his home and origin. There is no sabotage of time structure; the story is sequenced in the strictest, most straightforward way it could be.[2] Yet it startles and dismays us by the insight it brings into our time frame, the inexorable passing from birth or home to death or eternity, and the distortion of time in the perspective of our transient *now*.

It is more usual for a text to eschew this mathematical exactitude, and in any case, to leave the task of constructing the time frame of the story largely to the reader. Indeed, this construction is one of the reader's primary tasks as he goes along. Play with this imposed, ever-present task leads to structures that go beyond the imitation of memory—even for memories of prevision, previsions of the memory of prevision, and so forth, as paradigmatically exploited by Proust—to the invention of forms peculiar to texts.

One example, still relatively straightforward, is Thornton Wilder's *The Bridge of San Luis Rey.* We are told that this is a record made by a Franciscan, in eighteenth-century Peru, of the simultaneous deaths of five travelers. But we gain no acquaintance with this Franciscan, nor do we enter into his memory. The unity of the presented narratives lies in their simultaneous end. That final correlation is supplemented backward, so to speak, only in some fragmentary ways: the reader's assumptions of normalcy can latch onto the clues about approximate ages, incidental overlaps of the biographies, and echoes of events from these lives in each other.

Narratively more exciting is the subtle, innovative play with time in Ford Madox Ford's tetralogy, *Parade's End.* This already begins, though modestly, in the first volume's first paragraph: a simple past-tense description of two young men in a railway carriage, suddenly interrupted by the interpolated revelation "—Tietjens remembered thinking—." It was not really these events that were being related, but the memory of them. The main example, an episode that will continue to gain significance in retelling, comes in chapter 7. As we start reading the chapter, we are with Tietjens and Valentine on their long, all-night drive, but we do not know whether we are near its beginning or near its end. The timeless character that this long night has for them is reflected in the narrative dislocation in time and reinforced by the characters' dislocation in space. Every detail is perfectly clear, but vague in its location, so that even voices nearby come as if displaced by ventriloquists. Beginning, duration, and end—the when as well as the where—are removed from awareness, which has, of course, its powerful reasons to keep them at bay. Slowly the inevitable clues accumulate, the potentialities of the night condense, place-names appear, the dawn approaches, the reader retrospectively constitutes the duration and internal order of the night.

There is a similarly dislocated beginning for Ford's second volume, *No More Parades*. We are with Tietjens's wife, Sylvia, as she gets up from luncheon carrying her plate. The first volume has ended at the end of that long drive, still some years before World War I. In this new moment, in Sylvia's consciousness, references begin to appear to that war. We now relate the narrated events to our own history, to England and France. A reference to "the early days of the Great Struggle" surely moves us several years forward even from 1914? Suddenly, a dozen pages after the first line, Sylvia throws the entire contents of her plate; and we see Tietjens there in the room with her, for she throws it at him. He is in uniform. We have come five or six years since that drive in the country; we are in London, at war. While a salient example, this is not atypical of the narrative technique elsewhere: the scene begins, crystalline in every detail, but the moment is vague; then it slowly distills and becomes precise as memories and incidents appear and demand to be ordered, placed, located in time.

Today, of course, experimentation in narrative structure is no novelty; even the sabotage of the reader's task in, for example, the novels of Robbe-Grillet is—we now realize—but an exaggeration of the willful game author has always played with reader.[3] This was true even in those times when Aristotle's unities of time and place were taken to be definitive of effective literature. The reader correlates biographies of character and narrated events with each other and with his own history. But he does so always under the threat of reversal: the author's freedom to subvert later the expectations he has deliberately played on before is always part of the context of reading. Thus the correlation, the constitution of order in time, remains conditional and tentative until the end of the text. We have an ideal frame—we conceive of the events as in some definite order—but no rigid frame of reference. The locational function of each clue in the narrative, whether it refers to other parts of the narrated episodes or to episodes such as the World War in the readers' presumed common history, must thus remain fragile, equivocal, and undermined by the rights of future narration. Therefore the construction of narrative time is always essentially internal to the text, even when the text gives every sign of wanting to be related to extratextual reality.

The Relational Theory of Time

Perhaps some of you have come to this paper with the question whether or not time is real. Well, let me answer that right away. It is not—time is not real. Some philosophers disagree, but that is my view.

However, I mean this strictly and literally. Outside philosophy, people are not used to strict, literal speech. Ordinary discourse is much too poetical for philosophy. Some time ago, a telephone salesman called, and asked me brightly, "May I speak to Mrs. van Fraassen, please?" Now, I am not married. And I am a philosopher. So I answered him with the strictly literal truth. I said, "Mrs. van Fraassen does not exist." There was a ghastly silence for a moment; then he said "Oh . . . I am *so* sorry, . . . I didn't know. . . . "

Time is not real, time does not exist, there is no such thing as time. But events occur in some sort of order, some after others, some before, and some simultaneously. To me, that is no different from saying that an abstract entity such as the color spectrum is not one of the things in the world, nor is the *whale*, nor the fall of night—although there are colored things, which match or clash with each other, and individual whales, which give birth to other whales, and the paradoxical deepening and fading of colors at sunset. In this view about time, I oppose of course Newton's:

> Absolute, true, and mathematical time, of itself, and from its own nature flows equably without relation to anything external. . . . For times and spaces are, as it were, the places as well of themselves as of all other things. All things are placed in time as to order of succession; and in space as to order of situation.[4]

I also oppose the views of some of my contemporaries (for I would say the same about space-time as about time, *mutatis mutandis*).

Given this disagreement, however, I face a task: to explain how time order is constituted by means of, or on the basis of, relations between the events and processes to be so ordered. That is the proper task attempted under the banner of the *relational theory of time* (and space, and space-time).

At this point I'm sure you can see the intimate relation, for me, between narrative time and physical time. For I believe the constitution of time in our construction of the real world is not different in essential character from the constitution of time by the reader in his construction of the narrated world as he reads the text. I would like to discuss the principles of this construction briefly, under several headings.

Time as Causal Order

Leibniz outlined the first construction of time order from causal order:

> Given the existence of a multiplicity of concrete circumstances which are not mutually exclusive, we designate them as contemporaneous . . . we regard the events of past years as not coexisting with those of this year, because they are qualified by incompatible circumstances.
>
> *Time is the order of non-contemporaneous things.* It is thus the universal order of change in which we ignore the specific kind of changes that have occurred.
>
> When one of two non-contemporaneous elements contains the ground for the other, the former is regarded as the *antecedent,* and the latter as the *consequent.*[5]

This passage marks a turning point in the history of the theory of time, the point when the analysis first directed itself properly to time order and not merely duration. From our present point of view it is remarkable also in how closely it relates the constitution of physical time to that of time in narrative. Compare the passage that begins the Russian formalist analysis of narrative time structure:

> We may distinguish two major kinds of arrangements of these thematic elements: (1) that in which causal-temporal relationships exist between the thematic elements and (2) that in which the thematic elements are contemporaneous, or in which there is some shift of theme without internal exposition of the causal connections.[6]

Compatibility and Self-Identity

In outline, Leibniz's conception is the only possible alternative to an absolute conception of time. If time is not an independently real arena in which each event has its appointed place (as in a diary where the date appears at each entry), then the order of events must derive from their own characteristics and mutual relations. In any case, if Absolute Time were real, we'd still need to ask how an event is located at one instant rather than another. This is the perennial problem with postulated transcendent realities, that they promise to relieve us of such mundane work (as explicating the order of events in their own terms) but then are impossible to relate to the world they promised to set straight for us.

However, Leibniz's conception itself has many problems. At first sight, it is simple enough: use qualitative incompatibility to separate the events that cannot be contemporaneous, then rely on causal order to arrive at a sequential structure. Let us begin with the first idea—that of separation in time.

Compatible events might, unfortunately, still not be contemporaneous. The Prince recalls that the first messenger first returned on the tenth day, while they were pitching camp. Had the Prince and his retinue only just come to a stop, had they just begun sinking the first tent pole, were they fastening the last canopy strap? Perhaps the Prince does not remember—and nothing internal to these events *themselves* will give him the requisite clue. It appears that he would have to relate them to still other events, not recounted in the narrative, to arrive at a judgment of temporal separation.

Indeed, a judgment of relation already plays a role even when noncontemporaneity is inferred from incompatibility. That some messenger is absent is not incompatible with some messenger's having arrived—they are incompatible only if it is the same messenger. So two events are incompatible only if, *first,* incompatible qualities are involved but, *second,* these incompatibles inhere in the same subject. This second point requires a relation between these two events—*genidentity,* the relation of involving the same enduring subject—which is by no means derivative from merely qualitative characteristics.

This genidentity, or enduring identity over time, which was left tacit here, can certainly not be taken for granted at this level of discourse. Indeed, it is a great mystery, severely called into question by Hume, and since then the subject of many failed reductions and ultimately denials. Today such philosophers as David Lewis and Derek Parfit are well known for the denial of enduring self-identity over time; their doctrines are especially provocative when applied to persons.[7] There is an intermediate position that says there are only events (which are momentary states), but there is a special relation between them, again called "genidentity" but not definable in any way. What we call the history of an object is just a class of events connected by this special relation. On this position, Leibniz's construal of temporal separation still does not work, since there is no *logical* force to the claim that genidentical events involving incompatible properties must be nonsimultaneous.

When the identity of an object over time is clear, the incompatibility of ascribed characteristics is certainly a definitive clue to temporary separation. But even relatively clear instances of incompatibility give at best mixed indications of time order, especially since the incompatible characteristics ascribed are usually relational in some way themselves. A typical illustration is provided in the following passage from Gérard Genette, where the incompatible characteristics are the relational ones of *sleeping in* different rooms: "The first temporal section of the *Recherche* . . . evokes a moment that is impossible to date with precision but that takes place fairly late in the hero's life. . . . One of the rooms evoked is that of Tansonville, where Marcel slept only during the visit recounted at the end of *La Fugitive*."[8] It seems clear therefore that the relational theory of time will have to rely almost entirely on relations, and that merely qualitative aspects help constitute time order only very little.

The Causal Order

The passage I quoted from Leibniz continues, very revealingly, to spell out his hopes for causality as constitutive of time order: "My earlier state of existence contains the ground of the later. And since, because of the connection of all things, the earlier state in me also contains the earlier state of the other thing, it also contains the ground of the later state of the other thing, and is thereby prior to it." For this reason his version of the relational theory is called a causal theory of time.

But causality is a thoroughly theoretical notion. Is there really a fact of the matter? Or is causal structure our projection on the world? Some philosophers say the one, and some say the other. None says it is immediate and transparent to the intellect. It poses in any case the same problem as time itself. Either we postulate a relational structure that connects the events "from outside" as it were—a primitive relation that could in principle be there or not, independent of more mundane characteristics—or else we attempt to constitute the causal order on the basis of those more mundane characteristics and connections. If we do the former, we have not replaced Absolute Time with something more intelligible. If we do the latter, we are in effect continuing to construct the natural order without recourse to causality as such.

It is no accident that, in the above quotation from Leibniz, the reference is through a personal pronoun—but it is remarkable to find that reference in any essay on the foundations of geometry, time, and space. This conception of causality, which describes it as a relation of containment, indeed of logical implication, is quintessentially rationalist. It reveals—or should I say, betrays?—the heart of the great seventeenth-century unification in ontology of personal being and nature. But this synthesis was already disintegrating as Leibniz wrote. The conception of events, or states of affairs, as internally related did not survive in a natural philosophy oriented toward modern physics, at least not under the renewed onslaught of empiricism.

The difficulties are by no means a mere phantom of empiricist prejudice. In what sense could one physical event or circumstance contain the ground for another, how could there be an implication between events? This is precisely the point where the rupture between the modern view of material, physical reality and the categories of the personal becomes visible.[9] We can see such "internal" relations between events when the events are "intentional," that is, when they are acts characterizable essentially in terms of the intention involved. But we are unable to find intentionality in the world of physics—such characteristics are not physical characteristics. The regularities described by physics are simply the patterns in which events occur, as a matter of contingent fact.[10]

Consider Sartre's play *Dirty Hands*. There is no *logical* connection between the physical movement of Hugo's finger on the pistol trigger and Höderer's death. There is a logical connection between that death and Hugo's act of shooting Höderer on the orders of the Party. The relationship is still not one of implication, but there is now in the first act's description a reference, in the intention attributed to Hugo, to Höderer's foreseen death. This intentional language, entirely absent from modern physics, contains the paradigm for the rationalist view of causation.

At this point, if we are interested specifically in narrative time, we may be tempted to dismiss doubts about causality as relevant to natural philosophy only. For in literature, we are in the realm of intentionality. After all, given the intention that "governed" Hugo's act, we do have an answer to *why* Höderer died, the causal order is due to what the events themselves were like, and time order is derivative from that. But I chose the example carefully. Any doubt as to the why-because connection has been removed here by the intentional description of the events—but that doubt has not disappeared, it has simply been displaced. For it is entirely ambiguous, in *Dirty Hands,* whether the act was one of shooting on the orders of the Party. Did Höderer die because of his political acts and the Party's orders? The play raises the disquieting doubt that not only our knowledge but the facts themselves may fail to answer that. Certainly Hugo has no definitive access to an answer—and if he does not, then no one does. Is there a fact of the matter nevertheless? It would be sheer postulation to say so. The idea that causation is a clearly defined, objective structure at least in our actions and personal lives may derive merely from a philosophical wish: that those events are, on the one hand, internally connected by their intentionality and, on the other, as crisp and clear, definitely so or not so, as the events constituting the material world of modern physics.

The narrative text is thus as much an enigma for causality as the natural world. The "because" is conjectural, ambiguous; we are at a loss to find the objective demarcation between inference and fallacy, even in the most concrete examples—to such an extent that Roland Barthes, in just this respect, begins to equate interpretation and confusion:

> There is a strong presumption that the mainspring of the narrative activity is to be traced to that very confusion between consecutiveness and consequence, *what-comes-after* being read in a narrative and *what-is-caused-by*. Narrative would then be a systematic application of the logical fallacy denounced by scholasticism under the formula *post hoc, ergo propter hoc.*[11]

But when the basis of conjecture becomes so fluid, the very meaning comes into doubt (and not only for verificationists).

Correlation of Histories

The relational theory of time solved its problems in several ways, but mainly by minimizing the role of causality. When two events are connected causally they must at the very least be related either by genidentity (they involve the same body, or at least one persistent entity of some sort), or else they take part in some exchange that constitutes a signal, broadly speaking. But it is only through this minimal sense of connection that time order becomes definite—no further, more mysterious aspect of causation plays a role.

At least, that is the hope. These connections certainly establish sequences, and the sequences overlap because one event can belong to several such sequences. We can also imagine that many such connections are hidden, and would come to light only if investigated. But it is easy to see that in principle, great indefiniteness may remain. The facts *underdetermine* the time order, unless we include purported *modal* facts about what might, would, or could have happened.[12] To this curious ingression of the possible, we turn next.

Time and Possibility; Reality as Text

There is one great difference between reality and fiction. We all know what it is. You can demonstrate it, in imitation of the famous, by kicking a stone, or raising first one hand and then the other. The real is actual and the imagined merely possible. But if this difference is so great, why does it not engender other marks of difference, telling marks of solidity that will delimit and set off and distinguish?

No, this is the curious point, there could be no such mark. For if anything real bears any mark whatever, we can also imagine something (unreal) that bears that same mark. We can even imagine our fictions to be real or imaginary, as in Henry James's "The House on Jolly Corner," in which the apprehended ghost is after all not real but is—on one reading—merely the man the protagonist would have been if he had lived his life in New York City instead of London.

The Supposed Determinacy of the Real

There is just one way in which philosophers have persistently thought they could find a difference nevertheless. The real world is complete and settled in every respect; an imagined world is indeed *conceived of* as also complete in that way, but is not, for it is definite only to the extent specifically settled by the text. For example, each real Prince of Denmark has or has not had a mole on his left knee—and either has or has not owned an inky cloak. Though we do not know the details, reality has indeed settled it one way or the other. But Hamlet now—it is certainly implied in the text that he is a whole man and therefore had a left knee, and therefore either did or did not have a mole thereon. But this still leaves two possibilities, and the text does not settle that detail. And therefore it may be true of Hamlet that he was *thus-or-so,* but it is not true of him that he was *thus,* nor that he was *so.* Every text being finite in every sense, and with respect to details certainly exhaustible before conception is exhausted, we have here the great supposed difference between reality and what imagination creates.

This robust distinction received a rude shock with the advent of quantum mechanics, for most interpretations of this theory are at odds with the idea of the total determinacy of the real.[13] But the basic idea is not defeated, for there are constants in the theory that are always determinate in value, while an imagined world could be indefinite in any (logically nontrivial) respect. Earlier, the advent of relativity theory had brought a similar shock, for it denied that the question whether or not two spatially separate events are simultaneous has a determinate and univocal answer. However, the basic idea had been left intact then too, for the theory was immediately developed so as to give us a new catalogue of "absolute" relations, which remain invariant even if we change frame of reference. The earlier notion "simultaneous" had to be reconceived as elliptic for "simultaneous (under convention . . .) in frame . . . " The basic point remains: science conceives the world as determinate in myriad ways that go beyond our evidence.

However, so does every text. The world presented in a work of fiction is conceived as determinate in myriad ways going beyond the text itself. So we cannot conclude: science underwrites this philosophical explication of reality. We must only say: on that explication, what science describes is actually one way or another whenever the description says so, while what texts of fiction describe is not. This philosophical idea is in part about how to think of science, and therefore—though it has not been defeated by an inability to provide a reading of these new scientific texts—it clearly cannot be supported by them. We had better look a little further if we are to put it to the test.

Vagueness of Time Order

The indeterminacy of time order in narrative is pervasive, inevitable, and ineradicable.[14] Those characteristics are also typical hallmarks of triviality—what is more universal, and less remarkable, about texts than that they are composed of words? But in this case those characteristics mark a focus of significant narrative strategies as well. The reader expects a certain vagueness and takes it in stride, is quite willing to conceive of the events as determinately ordered somehow within the merely indicated outline—but the narrator, trading on this innocence, may return to those events later in his narration and reveal a hidden significance that undoes our compliance. And we, aware of the obligation of innocent compliance and also of the possibility of narrative subterfuge, read now with this tension of what is not yet settled in the text—which in turn allows the narrator to create his effects by mere feints, by the merest sidelong glance toward the past.

Sometimes the past is as it were nonchalantly revealed: Odette came from de Forcheville's house on the day of the cattleyas. Sometimes the narration reverses itself to the extent that a character loses some aspect of his past that so far served to define him for himself. To some extent this is true in *A Dance to the Music of Time,* when Nicholas, serving with Brent in the army, is forced to "rewrite" the story of his own affair with Jean Duport, though the task comes at a time when a good deal more has already entered his definition of himself. There is perhaps an added poignancy in the fact that the revision is made when it does not matter so much any more, an added pain felt at seeing how that part of his life has already lost so much to his indifference. This redefinition occurs as much in the reading as in the character since it occurs several volumes after the affair and thus after a long period of compliance by the reader. The redefinition of the past is much more radical in *The Death of Ivan Illich,* though there the revision, not preceded by narration of the past, is more perceived than experienced by the reader. Sometimes also the effect is gained rather from an ambiguity that is never removed. I have already referred to the evoked "moment . . . impossible to date with precision" at the beginning of the *Recherche*. Much more intricate and tantalizing is the incident with the little girl-cousin with whom Marcel experienced "for the first time the sweets of love"—she remains unnamed, the incident remains at an indeterminate time, datable only imperfectly by reference to his Aunt Léonie's domestic habits.[15]

Of course we are to conceive of the incident as having taken place at a determinate instant in Marcel's life. Yet there is no pretending that this narration, which came to an end when Proust ceased to write, still hides a detail that will settle the moment definitely. The question is now: how different is the situation in reality? Can we answer: but it *is* settled, one way or the other, though we don't know which way? Newton would have answered *yes* without hesitation. But how does the question fare on the relational theory of time?

The relational theory of time reappeared in various forms, from Leibniz through Kant, Whitehead, Russell, and Reichenbach to Grünbaum. The problem to what extent the facts determine the temporal order appeared clearly at a certain point in Russell. In Russell's variant, many problems are avoided by postulating that events are of finite extent and overlap quite literally. The overlap is chosen as the basic relation to define all other space-time relations. But will the defined structure be as we had conceived it? Will it be a continuum and ordered in the right way? The answer will depend on how many events there are and how they overlap. Accordingly Russell postulated appropriate answers to these prior questions. But are the postulates true? "Whether this is the case or not," Russell writes in his typically robust fashion, "is an empirical question; but if it is not, there is no reason to expect the time series to be" as we conceive of it.[16]

If there actually are only a few events ("few" as the mathematician speaks), we still conceive them as set in time in a determinate fashion. But the structure defined from their overlap relations will not be thus; it may have, for instance, only a partial and not a linear order. Now, Russell has obviously embraced the view that whatever structure is so definable from the actual event overlaps, that structure is time. If the definable structure is not like time was meant to be, so much the worse for time. But, Russell notwithstanding, the philosophical account of time should honor its conception, and failure to accord with that is failure for the account, not for the concept. When we conceive of a universe with few events, that does not change our idea—we do not conceive of that universe as having its own rather curious time—but we ascribe the discrepancy to the fragmentary character of the actual among the possible. There could have been more events, we think, and they would have filled out the usual structure all right. This is already what Leibniz had insisted on in a controversy with Locke: "The void which can be conceived in time, indicates, like that in space, that time and space apply as well to possible as to existing things. Time and space are of the nature of eternal truths which concern equally the possible and the existing."[17] But the problem remains in a different form: Suppose the actual relations among a paucity of events do not establish a determinate time order. Then *how* would the other possible events have made the order determinate, if they had occurred?

In the preceding section I outlined, in effect, a version of the causal theory of time that followed Reichenbach and Grünbaum. But a paucity of genidentity and signal connections is also conceivable, and would pose the same problem. For Reichenbach, we could put it graphically as follows: if "straight line" or "geodesic" has *light ray path* as physical correlate, what about straight lines in the dark? The actual physical correlates alone radically fail to determine geometric structure, in space, time, and space-time.

In the development of the theory, there was accordingly a shift from actual connection to connectability. The significant relation between two events is not whether they are actually connected but whether they are *connectable,* that is, whether some signal or trajectory, if it had emanated from the one, would have reached the other. But the assertion that time order is made definite by this relation relies on the tenet of *counterfactual definiteness*—that there is a real fact of the matter whether the signal *would* have reached if it *had* been emitted. The postulation of absolute time and its independently existing locations has been avoided at this stage only by postulating a solidity in the possible, by trading on counterfactuals and modality—a maneuver not unfamiliar in other regions of ontology, but of dubious value. Clearly we must take the discussion one stage further.

Reality as Text

In some respects, I am a very old-fashioned realist. There are trees and there are people; I see them, and they are there just as well when I do not look. There is no privileged observer of this reality, nor does observation have any logical connection with existence. Much is hidden from my sight, and from everyone else's as well; that does not make it unreal. But I would say this equally about a text. There is no privileged reader. A new reading aided by reflection on other texts or on experience, or merely by insight, may discover a structure theretofore unnoticed. Other readers may return to see it too. It would be utter sophistry, I think, to say that this structure was not already there, with equal objectivity, before it was uncovered. Since there is no privileged reading time either, we must perforce add that there must surely be many structures present in the texts we have, as yet hidden but to be uncovered later—or, by human frailty and finitude, hidden forever.

I have no hesitation in putting forward this view; but remember that I am a philosopher and so mean what I say literally, and that logic is mainly the art of showing just how little follows from our premises. This reality of which I have a robust sense need not itself be all that robust, solid, definite. That it has some hidden structure does not entail that *all* questions left unanswered on the surface are definitively settled in the depths!

When Leibniz insisted that time is the order of the possible as well as the actual, he must have realized that this insistence went much more readily with an absolute than with a relational view of time. Yet he continued to deny that time is real. Time, he said rather obscurely, is an ideal entity.[18] Let me say it in a way that is closer to us: time is a logical space. The color spectrum is the logical space for colored things. There is no need to add: and the color spectrum exists, eternally and at peace, in its own transcendent beauty, and so forth. Porphyry's tree is the logical space of everything, the Library of Congress classification is the logical space of all books. We conceive of any possible extended opaque object as determinately located in the color spectrum and of any possible book that might eventually be published as somehow placed in our beloved LC classification. None of this swells our ontology.

But it does not follow that the time order of real events is definite any more than that of narrated events in the *Recherche*. For although books are individually located in LC, only the structure of all events taken as a whole is set in time, since correct "placing" of events is determined by their mutual relations. And there may remain in principle more than one way to determine the placing. Rather than postulate some transcendent criterion of correctness—whether through counterfactual facts or in any other way—I suggest we accept the same imminent vagueness for the order of real events, underdetermined by the facts, as we do for the order of narrated events, underdetermined by the text. In both cases, the world is conceived of as determinate, but the necessity in how things are to be conceived does not engender a necessity in how they must be.

In this I have, I suppose, given up that demarcation between the real and the imagined worlds in terms of determinacy. In the philosophy of science, when a particular tenet of determinacy disappears, one can always add: but the world is still determinate in all the ways in which physics continues to conceive of it as determinate. In that way the demarcation would be maintained: world and texts would remain fundamentally different. We can indeed always add something like that, but we need not, and I see no loss if we do not. The stones we really live with we can still always kick, our real hands we can always hold up to show. The glory and the terror of this world remain, even if *this* text is not so different in kind from other texts.

POSTSCRIPT CONCERNING REALITY AND FICTION

Reality as Text, Continued

The criterion of demarcation between fictional worlds and the real world, which we examined and found wanting, focused on a certain sort of finitude. In the case of time, we found a parallel between the two worlds: each is conceived of as totally definite but constituted very incompletely by the actual text—which consists of written words in the one case, and of all physical events in the other.

There are other ways in which the finitude of the literary text might distinguish it from reality. Here is a second attempt at demarcation: The real world is not constituted definitely in a single observation or by a single observer. But a single reader can take in, observe, the entire literary text—the entire basis on which the constituted fictional world is erected. Is that not a difference?

There are two objections to this. The first concerns the problem of the definitive text. The second concerns what the reader brings from outside the text, and the extent to which that contribution may have a legitimate, inalienable place in the constitution of the fictional world. If there is no definitive text or if the text is augmented in a certain way by every generation of readers, the cases of fiction and reality are to be seen as parallel in this respect also.

The Question of the Definitive Text

It is certainly not guaranteed that I, one reader of *Parade's End,* have taken in the entire text. In the case of more venerable works, we are well aware of the fragility of historical judgments that declared a certain version definitive. One entire novel, Umberto Eco's *The Name of the Rose,* was written on the question of what the text of Aristotle's *Poetics* really is. We take it that this text lies scattered in bits throughout history—we have access to two medieval Greek manuscripts and two medieval translations, one into Arabic and one into Latin. There is a one-page remnant of a Syrian intermediary between some Greek manuscript and the Arabic. These sources disagree. All are of the first book of the *Poetics* only. With respect to the remainder, the most we have is a medieval manuscript now in Paris argued to be a summary of the second book, about catharsis and comedy.[19]

But if we do not *have* a definitive text, surely there *is* one, namely the physical inscriptions that actually left Aristotle's pen? They are lost, destroyed, but so are all past events in reality; the difference is that after Aristotle's death, what had left his pen is the complete text *ipso facto,* and creation has ended. No death or passing ends the growth of reality—creation continues.

This difference is challenged, however, by the query concerning what exactly the text is. We could arbitrarily end the definition of that text at the end of Aristotle's fingertips, but can we do so with reason? Let us try this general idea: the text is the factual basis for the constitution of the depicted world. Now it is clear that each reader constitutes that world for him or herself, and that this construction introduces many peculiarities that are to be ignored or classified as idiosyncratic. But not all introduced differences are idiosyncratic. A reading is communicated, and scholars communicate the results of such communication, what is agreed to be common and what is not. The next reading is in the light of, and is colored by, this inheritance, this social process, the residue of previous readings. That residue also includes a record in which the narrated events are related to common history and to other literature that postdates the author's death.

The point is, we cannot very well insist on the one hand that the world created by or in the text is never definitively constituted while holding on the other hand that the factual basis for that constitution is unchanging. For the validity of reading the old text in the light of previous readings, for example, elevates the reports of previous readings to new elements of the text. This is so even if "in the light of" means in part "while classifying certain previous readings as incorrect vis-à-vis the text in respect to so-and-so."

Borges has the habit of ambiguity: his irony often consists in a refusal to reveal, avow, or disavow irony. So I can cite "Pierre Menard, Author of the *Quixote*" as both in support and in detraction of the previous paragraph. But Borges's irony would be flaccid, ineffectual, and dead for us if it did not trade on insight and truth. I take it that it is our reluctant, even resentful surrender to this insight—that Cervantes did not end the process of relevant accretion to the text of *Don Quixote*—that gives the story its force.

The Possibility of Idealism

Here is a third and final attempt at demarcation. Gideon Rosen made it in the following summary form: "Idealism is right for fiction and wrong for the world."

When say "we" or "us" in this philosophical context, we refer to a whole community of persons, of rational beings, stretched out in space and time, past, present, and to come. In the world of the *Recherche* or *A Dance to the Music of Time* there are structures that we shall later agree *we* had not yet uncovered in A.D. 1988. But are there in this fictional world any structures that it is not within our power or capacity to uncover—that we could not uncover in the future, regardless of how their readership evolves? The fact that this work is a humanly created text—even if not definitively determined by its author alone—makes it impossible to answer that question in the affirmative. Whatever real structure it has, even in the fullness of time, is there because we shall have given it, and hence is accessible to us.

But it is possible to say *yes* to that question if asked about the real world. It is possible that there are aspects of this world we *cannot* ever uncover, which transcend all powers of observation, detection, verification—perhaps even our capacities for theory and conjecture. So it is possible not to be an idealist about the real world, to be a realist (in the sense just described). But it is impossible not to take the idealist position about the fictional world.

Notice that I have amended the assertion in effect from "it is so" to "it is possible for a philosopher to hold." If realism is the position that there are transcendent structures, then realism may be false. But if it is possible for there to be such structures and there are none, we shall also not be able to find that out. So this final demarcation does not point to any definite difference between real and fictional worlds; only to a difference in what one can *tenably* assert about them. Real and unreal are finally distinguished only by the activity of the philosopher!

But that is not so. Certainly, this criterion in terms of tenable philosophical positions is correct. It is very abstract, and I see no less abstract criterion to formulate. Nevertheless, there also remains the difference we cannot formulate but only show: I can only touch *this* flesh, *these* stones, *this* wood.

Time and Creation

CORNELIUS CASTORIADIS

I

WHEN THINKING ABOUT time—as, indeed, about anything—we cannot avoid encountering immediately an unsurmountable division:

- Time for us—or for some subject, or being-for-itself, with various evident and, at the same time, enigmatic characteristics (be it only its pulverization among all actual and possible subjects);
- Time in or of the world, as receptacle and dimension of whatever may appear, and as order and measure of this appearance.

Let us call these, provisionally, subjective and objective time.

Immediately, then, arises the question of time *as such,* a third term making it possible to talk about subjective and objective times as *times.* Time as such would then appear as *overarching* not only the various subjective times—mine, yours, the time of the Aztecs and the time of the Westerners, the time of the whales and the time of the bees—that is, the varieties of private times or times for a subject, but also all particular times of whatever nature, including objective time and its possible fragmentations (there is such a fragmentation in general relativity), and making possible, through innumerable articulations and encasings, their mutual adjustment, or at least accommodation and "correspondence."

So, we do speak, can and have to speak, of time in general, but must always bear in mind that there are many species of time—or many meanings of the term, in the same way as Aristotle used to say of being that it is a *pollachōs legomenon,* a term used in many different ways. This mention of Aristotle is not accidental: I shall argue that time is inseparable from being. It is not just that we happen to give various meanings to the same term, time, but also that *there are* different categories of Time. Why are they categories *of time:* that is, what do they share in common or, to put it in a more radical form, why is there a unity and unicity of time, if indeed there is one—these are questions to which only a very complex attempt at an answer is possible. Here again, the situation is the same as the one relative to being and, if I am correct, for the same deep reasons.

I spoke about subjective and objective time. Why take over and endorse this old-fashioned and platitudinous distinction between the subjective and the objective? I will return to this question later. For the moment, I will assert that there is being as subject, or that there are beings which are subjects, that is, are for-themselves. For instance: *we*. Now a subject is nothing unless it is the creation of a world for-itself in a relative closure. This world (receptacles, elements, relations, etc.) is what it is *for* the subject, and would not be as such and as it is unless there were a subject and *this* subject (and/or this class of subjects, etc.). This creation is always creation of a multiplicity. This we just find and state; we can neither deduce nor produce it. This multiplicity is always deployed in two modes: the mode of the simply different, as difference, repetition, ensemblistic-identitary multiplicity (for brevity: ensidic multiplicity); and the mode of the other, as otherness, emergence, creative, imaginary, or poietic multiplicity.[1]

But I shall also assert that—unless one is to give oneself fully to an absolute solipsistic delirium—subject does not exhaust being. First of all, which subject? There are indefinitely many subjects and modes of subjective being; and there is no way I can construct the existing and effective organization and functioning for-itself of a crocodile or a beehive as a product of my (or the transcendental) consciousness. Neither can I forget that the world of the beehive entails necessarily the world of flowering plants; or—to stop here an unlimited series of inferences—that the world of plants has to do with some properties of, or possibilities supplied by, inorganic matter. To be sure, whatever I say about all this is also in a decisive way co-determined, co-organized, by me as subject. But—and we will revert to this argument—whether I think about the organization I, as thinking subject, impose on whatever there is, or about the organization which living beings in general both present and impose on their world, it remains that neither would exist if the world as such, in itself, were not *organizable*. Subjects cannot exist outside a world, nor in any conceivable world. The meaning of the term "objective" is, here: the possibility supplied to subjects as beings for-themselves by what there is to exist in a world and to organize, each time in another way, what there is.

Two consequences follow.

First, we can never separate rigorously and absolutely or ultimately, in whatever we say about the world, the subjective and the objective component.

Second, we cannot restrict the two modes of difference and alterity to the world of the subject(s): they are inherent in the world as such. There is ensidic multiplicity, difference, repetition: there are tree*s,* in the plural; a cow produces calves and not parrots, etc. And there is creative, poietic multiplicity: a jaguar is other than a neutron star, a composer is other than a singing bird.

These two implications—as well as, most of the time, the premises underlying them—have been on the whole neglected or ignored by inherited philosophy. And this in two ways rather than one.

First, on an abstract level, inherited philosophy works with a radical separation of subject and object. The result has been a hovering between a subjective and an objective position (*Einstellung*). This does not change, but is rather brought to the extremes, when the one absorbs the other—as happens with idealism or materialism. This takes place, of course, on the gnoseological (*Erkenntnistheoretisches*) level, with empiricism (inductivism), and apriorism. What is neglected in both views is that there is a subject and that any world is a world for a subject—whilst, at the same time, there could not be a subject and a world for this subject without a world which lends itself to the existence of subjects and to their knowing something (this boils down to the same thing) about the world. Whether we consider the most naive "perceptual faith" (Husserl, Merleau-Ponty) or the most sophisticated philosophical or scientific thought, we are always in the middle of a subjective world which would not exist, nor be the way it is, simply because the subject is what it is. But the same is true on the ontological level, when the essential determination of subjective being, its self-creation *qua* subjective being, is either denied to it—or denied to the being of the world, considered as an inert assembly of elements subject to perennial, self-identical determinations.

Second, on a more concrete but no less fundamental level, with the simple polarization or separation between subject and object, philosophy has ignored the *social-historical,* both as proper domain and mode of being, and as the *de jure* and *de facto* ground and medium for any thought. This can be seen in the way philosophy, from Plato to Heidegger, has structured its domain. This it has done by the positing of a polarized couple: the subject or ego on the one hand (psyche, animus, transcendental consciousness, ego, *Dasein* as the *je eigenes, je meines*); the object or world on the other hand (cosmos, creation, nature, transcendence, *Welt* and/or

Being). What is covered up in this way, never in fact thematized and never understood in its proper philosophical weight and character—its character as the condition, medium, purveyor of forms and active co-operant in any process of thought—is the social-historical, which is always, both *de facto* and *de jure,* the co-subject and co-object of thought. The actual, concrete embodiment of thought is, of course, the thinking, self-reflective subjectivity—but this subjectivity is, itself, a social-historical creation.

The twin results of this occultation have been:

- □ That the subjectivity about which philosophy talks is always a bastard construct, combining in various proportions elements of the psychical proper, of the socially-historically instituted understanding and reason, and of the self-reflecting activity of the social individual at a certain stage of history;
- □ That the world (or being) considered by philosophy is thought irrespective of its social-historical construction (i.e., creation), with the results, *inter alia,* that (1) the true question of the world as the ground for all the various social-historical creations of it is covered up, and (2) the deeply historical character of knowledge, the existence of a genuine history of knowledge, is either made impossible (Kant), ignored, reduced to a moment in the "forgetting of Being" (Heidegger), or downgraded to a sheer relativistic, "sociological" or other version—eliminating, in all cases, the question of the history of truth and of truth as history.

It is beyond my purpose and terms of reference to deal here with this inherited situation for itself. I will only exemplify and discuss it briefly at the level of the question of time.

As Paul Ricoeur has correctly shown (in *Time and Narrative,* especially Vol. III) philosophy has always dealt either with subjective or phenomenological time (Augustine, Husserl, in essence Heidegger) or with objective or cosmological time (Plato, Aristotle, Heidegger's "vulgar" understanding of time; Ricoeur only in passing mentions Plato and places in this category also Kant, a decision raising questions which I will not discuss here), with the result that each advance in the understanding of the one only multiplied the difficulties in understanding the other as well as in bridging somehow or other the gap between them.

Now this division can easily be understood and interpreted on the basis of what has been said above. Philosophy has concentrated:

- □ Either on a reified, identitary (ensemblistic-identitary, *ensidic*) time, which would form, supposedly, the backbone of physical experience and has as such to be essentially measurable, therefore is conceived centrally from the point of view of the repetition of the identical (periodicity, etc.), ignoring, among many other riddles, the basic datum of the emergence of Otherness;

□ Or on an experienced, lived time, which as such can only be, in each case, utterly subjective in the derogatory sense of the word (*je eigene, je meine,* says Heidegger) and makes of the existence of both a public and a cosmic time either an intractable aporia or the outcome of the fall of the subject (the *Dasein*) into everydayness and inauthenticity and of its forgetting Being and covering it up with simply "encountered" particular being (*Vorhandenes*).

Let me illustrate, at this point, the fateful effects of the covering up of the social-historical. If the social-historical had been, as it ought to, placed at the starting point of the reflection, a part of the aporias of time would have been dissolved, and another part brought under another light. One would have immediately perceived both the solidarity and the distinction between identitary and imaginary time; the necessary leaning of the former on the first natural stratum (therefore on a cosmic time); the irreducibility of imaginary time to identitary time, but also their inevitable interpenetration; the fundamental alterity of the imaginary times instituted by societies which are other, opposed to the relative homogeneity and commensurability of their identitary times taken as such (abstractly and in separation). One would also have seen that each society, as being for-itself, entails the creation of a proper (imaginary) time, consubstantial to its being-thus (being a society and this particular society).

II

To illustrate the aporias engendered by the objective (or cosmological) and subjective (or phenomenological) approaches a short discussion of two eminent proponents of the respective views is useful.

Leaving aside Plato—in the *Timaeus* (37d), time is clearly posited as an identitary, objective, measurable ordering of everything worldly—we find in Aristotle the first systematic and thorough exposition of the objective, cosmological point of view. The well-known locus is *Physics,* Book IV, Chapters 10–14 (217b 29–224a 17). Without going through the intricacies and the extraordinary richness, subtlety, and solidity of the argument, we pluck the solution given by Aristotle, as usual, in the canonical form of a definition:

"Time is the number [numbered number, measure] of movement according to the before and after" (219b 1–2; 220a 24–25). Movement for Aristotle, we must remember, is not only local movement, but change in general (and this is reaffirmed in several places in *Physics* IV). Time *is* not the change (movement)—but it is one of the essential determinations of the change (movement). And it is also one of the essential determinations of change (movement) to be measurable.

Let us grant this. We cannot, though, help asking: what is the "before and after"? The explanation of it (219a 10–25) betrays a slippage papered over by the harmonizing interpreters and commentators of Aristotle. Despite the repeated, in *Physics* and everywhere, metaphysical and physical thesis that local movement is but one of the species of change (*metabolē;* hinted already at *Ph.* IV, 10, 218b 19–20), which comprises, beyond local movement, generation and corruption, alteration, and increase/decrease, that is, changes according to essence, to quality, to quantity, and to place (*Ph.* VIII, 7, 261a 27–36), Aristotle asserts, here and elsewhere (219a 11–25; cf. *Ph.* VIII, 7, 261a 26–27), that the local movement is the "first" (in the sense of the most important) and that "the before and after" is, firstly (originarily), in the *topos*—place, location, space. We would thus have to take "the before and after" as a spatial ordering—the spatial before and after of a moving body—which the temporal ordering follows (*akolouthein,* 219a 19), since movement (locally defined) and time always accompany each other (ibid.). But any spatial ordering is, of necessity, arbitrary. That for Aristotle this arbitrariness is not absolute, the Earth having a privileged or rather unique position, is irrelevant. We do not necessarily measure movements in relation to the center of the Earth.) A subjective element, therefore, inevitably creeps into Aristotle's cosmological view of time; and this is manifest in the formulations of *Physics* IV, 11: "we take cognizance of time, when we have defined the movement by defining the before and after; and only then we say that time has been [has elapsed] when we perceive the before and after in the movement. . . . for, when we think [*noēsōmen*] that the extremities are other than the middle, and the soul pronounces the present/instants [*nun*] to be two, the one before, the other after, it is only then that we say that this is time" (219a 22–25; 26–29). The after and before becomes thus a primitive notion, the understanding of which must appeal to some subjective ordering by the soul (the observer). I shall return to this shortly.

Seven centuries later—and leaving aside the very important Stoics and Plotinus—we find in Augustine (*Confessions, XI*) both the first clear formulation of the subjective approach, and a rebuttal of the conception of "a philosopher" which, very probably, he takes to be the conception of Aristotle, whom, equally probably, he has not read—or, if read, not understood. I start with the latter. Time, says Augustine, cannot be the movement; for we see the same movement taking place with different durations (*XI,* XXIII, 33–40; XXIV, 1–5). The argument of course has nothing to do with Aristotle's definition. Aristotle did not write that time *was* the movement (he wrote explicitly the contrary), but that time was one of the essential determinations of the movement, i.e., its *measure*. If "the same movement" takes place with different durations, then it is simply not the same movement, since one of its main determinations is its temporal measure. Nobody in his right senses would suppose that Aristotle did not know the difference between walking home and running home—or between the tortoise and Achilles. But also Augustine's misplaced rebuttal conceals the central aporia of his own position (and that is why it is discussed here): how does he know that the two "different durations" of "the same movement" are different, except by comparing them with a *tertium quid*—e.g., another movement supposed to run at a constant rate during the same "time"? "The same movement" here can only mean: the same spatial end-points, and thus the argument does not make sense. Augustine goes on to say, *Non est ergo tempus corporis motus* (XXIV, 29), the time therefore is not the movement of the (a) body, and that we measure movement as well as rest by *nostra dimensio,* our measure. But could we measure rest if everything were at rest? An Aristotelian would of course remark that we "measure" rest as the time during which another measured movement took place, i.e., by reference to and comparison with movement.

But, if Augustine's dialectics are poor, his central intuition is strong. We can measure time, he says, because there is a *distentio* (*distendo:* to extend, to deploy)—an extension or tension or deployment. *Distentio* of what? *Distentio animi*—a stretching of the mind. *In te anime meus tempora metior:* I measure the times in you, my spirit (or mind) (*XI,* XXVII, 26). And I measure this *distentio* insofar as *aliquid in memoria mea metior quod infixum manet*—insofar as I measure something that remains fixed in my memory (ibid., 24–25).

Thus, time is strictly correlated to the capacity or possibility of the mind to measure the affection (or impression: *affectionem*) *quam res praetereuntes . . . faciunt . . . et manet, ipsa metior presentem:* it is this very impression made by previous things, and which remains, that I measure as present (or because it is—still—present) (XVII, 4–6). Time is, in fact, the *distentio* the *animus* lives through and that remains (in the memory) and, therefore, can be re-called, called back unaltered—unaltered at least *qua distentio.* (Insofar as measure, in the proper sense, is concerned, this is obviously an untenable position; we will revert to this.) Thus the future, the not-yet (*nondum*) "consumes itself into the past" and the past, the no-more, grows (*crescit*) because the *animus* is capable of three activities or postures: *et expectet, et adtendit, et meminit,* it expects, it pays attention to (or attends to—is preoccupied with, Heidegger!), it remembers (XXVIII, 1–6). So that (or: in view of, *ut*) "the expected through that which is attended passes into [*transeat*] that which is remembered." The *animus*—the mind—is in time and/or makes time be insofar as it is *distentio* uniting these three "moments," expectation, attention, memory. If it is capable also of measuring time, it is because of this strange possibility of quantification supplied by the accretion of memories in memory.

Then Augustine reverts (L.XII) to the question which had, to begin with, put him on this treacherous track and remains the motor of his enquiry. "What was God doing before the creation," this silly and blasphemous question, is answered in principle by the distinction between eternity, as *nunc stans,* immutable now, and time, which belongs only to the created. But if time is linked to the created *animus* not only for its measure or its perception, but in essence, as the developments of L.XI tend to show, intractable difficulties arise, which will lead Augustine, in the XIIth Book of the *Confessions,* to a flagrant contradiction, as we will shortly see. In the meantime, let us note the decisive influence of Augustine on the time conceptions of Kant, Husserl, and Heidegger.

We discuss Augustine to exemplify a basic position and its aporias. Let us take up one of these (common to all subjective approaches). Augustine does not (and could not!) simply say: there is time only insofar as there is an *animus* and a *distentio animi.* He says, as we have seen, and has to say, I measure time by this *distentio.* This creates an impossibility, shared by all subjective approaches to time. How could a *distentio animi* supply a common, public time, and a common measure of time? How could it even supply a measure, in the proper sense (one has to suppose that Augustine knows what *metior,* "I measure," means), of private, subjective, personal time? Augustine's referents are purely subjective (even if they were taken to be non-psychological): expectation, attention, memory. In order to arrive at a common time and a common measure of time, all *animi* would have to be endowed *a priori* not only with an abstract capacity of measuring time, but with the capacity to measure the same time and strictly in the same way. Let it be said in passing: in this perspective, the existence for all subjects of a flow of time with the same direction has to be taken as a sheer fact, not susceptible of further elaboration or elucidation. The structure of subjectivity is such that it lives in attending, expecting, and remembering (in Husserl: attention, retention, protention; in Heidegger: the expected, the memory, the present), and with the same ordering of events for all. The concrete content of this ordering for me must be the same concrete ordering for you. This seems self-evident to Augustine (as, indeed, to all philosophers), so he does not even mention the problem. But also: the measuring has to be done with the same yardsticks and lead to the same results, without any external *repères*—bearings and benchmarks. Thus all *animi*—or the *animus* as such—must be such that their measuring operations are identical, in a way wholly independent of any "quantity" and "quality" of the memories stocked in each case. Then why refer to these memories at all? And why address always the *animus* as *animus meus,* as Augustine does (this will become the *je meine, je einige* of Heidegger)? But even in the case of the single *animus:* what is there to ensure the identity of successive measures of the same—or the comparability, as to measure, of different stretches of time—each one obviously filled with different memories?

Let me allow myself an aside at this point. Do we die *after* we were born because we expect death and remember birth (Heidegger's *Geworfenheit* and *Tod*)? A sentence absurd on the whole—and wrong or non-rigorous in its second part. We do not remember birth; strictly speaking, we do not know properly, *eigentlich,* that we were born. The *Dasein* does not know it was born; it only has been told so, and has seen other people being born. Neither does the *Dasein* know *eigentlich* that it will die; it has been told so, and has seen other people die. Nothing in me, nothing *meines* and *eigenes,* tells me that I was born and that I will die—nothing "psychological," and nothing "transcendental." That I was born and that I will die is essentially *social* knowledge, transmitted to/imposed upon me (and which, of course, the innermost core of the psyche simply ignores).

We revert to the aporias of the subjective approach. In Augustine's theological framework, the difficulties could be accommodated by divine construction (what could not?). But this way is barred for a philosopher (thus also for Husserl, Heidegger, etc.). And, even for the theologian Augustine, the way remains fraught with enigmas, since he has clearly based his whole argument on strictly subjective notions (memories, etc.). Therefore equivalent *a priori* properties of the subjects have to be postulated *ad hoc.* The question is important because, in another framework, it persists through Kant up to Husserl and Heidegger, where it becomes intractable.

It is useful to show this in the case of Kant. For Kant, time as a pure *a priori* form of intuition forces so to speak whatever appears, external as well as internal, into one single dimension of succession. The application of this form to whatever appears (the phenomena) requires the mediation of a transcendental schema, supplied by the transcendental imagination.

This schema is the "line." We have here, obviously, a shift of the problematic of time toward the problematic of space. But even this shift will not do. A fundamental property of time (any sort of time) is irreversibility, and there is nothing irreversible about a line: the total order on the open interval (x,y) is isomorphic to the total order on the open interval (y,x). Then, also, time is and has to be measured. In the case of space, one can admit measuring without any external support (as this is done, e.g., in mathematics): pure intuition compares segments, finds them equal or unequal, and so forth. But this presupposes that segments can be superimposed on each other or made congruent (in pure intuition). But segments of the "time line," by their very nature, are not superimposable. How can then they be compared in a valid way, and how could time be measured? Without something inherent in the phenomena as such, and which cannot be supplied by the transcendental subject, that is, without the existence of the effective repetition of equivalent pairs of phenomenal occurrences for which it can be rationally postulated that they are separated by equivalent intervals, there can be no measure of time—and no physical experience (*Erfahrung*) in the sense of Kant.

Now it is important to observe that neither Aristotle nor Augustine can hold his position to the end.

I gave above two quotes of Aristotle which link, in an ambiguous way, time with the activity of the soul (*Ph.* IV, 11, 219a 22–25, 26–29). But there is more. If nothing changes in our mind, or if the change escapes our notice, it seems to us (*dokei*) that no time has elapsed. But if there is "a movement in the soul," even if we are in the dark and nothing affects our body, it seems to us immediately that time has elapsed (*Ph.* IV, 11, 218b 21–219a 2 and 219a 4–6). So, the soul cannot perceive time unless there is *for it* a change; but also it can produce itself the change (the "movement") through which time is given to it. And when, toward the end of his enquiry (ibid., 14, 223a 25–26), Aristotle discusses the aporia Would there be time, if there was not soul?, his answer raises more difficulties than the interpreters would admit. If there is not a numbering subject, says he, in explaining the aporia, there can be no number (and time is the number of movement, let us remember). Therefore "if nothing else has in its nature [*pephuken*] the possibility to measure except soul and the mind of the soul, it is impossible for time to be if the soul is not, except that which is the substratum of time [*ho pote on*], in the same way as it would be possible for movement to be without the soul." In a soulless *kosmos* there would be movement, because *phusis* is movement; there would also be the "substratum" of time; there would be no time in the full sense—in the sense given by Aristotle's own definition—since this requires the "numbering" or "measuring." The difficulties in this solution are linked with the more general difficulties of the fundamental Aristotelian distinction between potentiality and actuality, which cannot be discussed here. Suffice it to conclude that for Aristotle (a) the soul itself can produce a substratum for time through its own movement (no wonder in this, since *psuchē* either is *phusis* or is strongly linked to it), and (b) the actualization of time as measure of the movement entails the activity of the soul.

Things are simpler with Augustine, who contradicts himself openly and naively. Further on in the *Confessions,* when the discussion of the Creation is resumed, he asserts flatly: "You [God] have from this quasi-nothing [*paene nihilo:* the initial *informitas* created before everything else] created all these things in which this variable [*mutabilis*] world subsists and subsists not, where the mutability itself appears, in which the time can be perceived [*sentiri*] and measured, for the times are made through the mutations of things, when the appearances [*species*] vary and change [*vertuntur*]" (*XII,* VIII, 28–32; cf. ibid., IX, 13–15). But he had already written (*XI,* XI, 4–10): "whoever would say to me that, if all appearances were suppressed and annihilated and if only the *informitas* [form-lessness] were to remain, through which everything changes and varies from one form to another, this form-lessness would exhibit the vicissitudes of time? This is absolutely impossible, for there are no times without the variety of motions and where there is no form, there is no variety." Time here has ceased to be just the *distentio animi,* the stretching of my mind; it is that in which the forms *vertuntur,* are changing into one another, and it is produced by this mutation of forms, strictly dependent on it. In other words, God has to create time together with his giving form to the initial, itself God-created *informitas*. (That the whole is a paraphrasis in Old Testament dress of Plato's *Timaeus* is obvious.) And this is not a *lapsus* of Augustine: he has to say that, since the Creation is a story unfolding in time (in cosmic time, from the *diesseits* point of view), in which the creation of the soul comes *last*.

III

Time belongs to any subject—any being-for-itself. It is a form of the self-deployment of each being-for-itself. Being for-itself (e.g., any living being) is creation of an interior, that is of an own world, a world organized in and through own, or proper, time (*Eigenzeit*). This is neither a deduction, nor an explanation. We take it to be a fact requiring elucidation.

This is most evident for us, as a primary datum, in the case of the human psyche—unconscious as well as conscious. To be sure, we have also to reckon here with the fact that, in a deep sense, the time of the unconscious and the time of the conscious are definitely not "the same"—though they act upon each other. But this would be an object of enquiry in its own right—as would, e.g., the time of a psychoanalytic session and the time of the treatment.

Psyche is irreducible, in its kernel, to society. The true polarity is not between individual and society, but between psyche and society. The individual is a social fabrication. But psyche cannot survive unless it undergoes the process of socialization, which imprints on it, or builds around it, the successive layers of what, in its outer face, is the individual. Socialization is the work of the institution, mediated of course in each case through already socialized individuals.

In and through the process of socialization the psyche absorbs or internalizes the time instituted by the given society. It henceforth knows public time—and has to go on living, coping with the difficult cohabitation of the various layers of its own, private time with instituted, public time. The same is of course true for all the rest: ideas, cathected objects, etc. The difficulty, or rather the clash, manifests itself not only in the opposition between the finite horizon within which the private time of the individual has to be lived (death) and the indefinite social horizon of time, but also, and equally importantly, in the difference between the rhythm and the quality (both extremely variable) of private time, and the steadiness, fixity, and prearranged variations in quality of public time.

Society (societies as such) is a type of being-for-itself. It creates, in each case, its own world, the world of social imaginary significations embodied in its particular institutions. This world—and such is the case for all worlds created by a being-for-itself—appears as the deployment of two receptacles, social space and social time, filled with objects organized according to relations, etc., and vested with meaning. Why receptacles, and why two receptacles; how far can these receptacles be separated from what they receive and from the subject to which they appear as receptacles: these are the questions ultimately reflecting the multiplicity of being, to which I shall revert later.

Descriptively, we always find social (public) time (and space) instituted in two intertwined threads.

There is and always has to be identitary (ensidic) time, the backbone of which is calendar time, establishing common, public benchmarks and durations, roughly measurable and characterized essentially by repetition, recurrence, equivalence.

But social time is and always has to be also, and more importantly, imaginary time. Time is never instituted as a purely neutral medium of or receptacle for external coordination of activities. Time is always endowed with meaning. Imaginary time is significant time and the time of signification. This manifests itself in the significance of the scansions imposed on calendar time (recurrence of privileged points: feasts, rituals, anniversaries, etc.), in the instauration of essentially imaginary bounds or limit-points for time as a whole, and in the imaginary significance with which time as a whole is vested by each society. There is the time of the perpetually recurring return of the ancestors; of the inner-worldly avatars of human souls; the time of Fall, Trial, and Salvation; or, as in modern societies, the time of "indefinite progress." Imaginary time is constituted inseparably from the three dermas (as I would like to call them, borrowing a term from embryology), the overlapping, overfolding, and interpenetration of which weaves together society: socially instituted representations, affects, and drives (pushes). The link of imaginary time not only with the creation of a social representation of the world strictly speaking, but with the fundamental drives of a society and its fundamental affects (*Stimmungen,* moods), is obvious, but would bear lengthy examination. Thucydides (I, 78), describing and opposing the moods and behavior of Athenians and Lacedaemonians, shows clearly their intimate link with the manner each of these two societies lived time.

This creation of (social) time by society requires in and of itself an elucidation. But another aspect has to be stressed first. Society always leans, and has to lean, on the first natural stratum insofar as the identitary (ensidic) dimension of its creations is concerned. (And this is true for every being-for-itself.) I dare think that a proper consideration of this fact forces us to admit the two cardinal and, to my mind, self-evident propositions which I mentioned at the beginning of this text and which ought to end the perennial philosophical dispute between objectivism and subjectivism and place the question these conceptions have tried to answer on a new ground:

□ Societies know—as do the fox and the hedgehog—at least something of the world. (Otherwise, they could not exist.) But this entails that, at least in some of its aspects, the world is know*able,* lends itself to some knowledge (whether empirical, relative, etc., is totally irrelevant for the present discussion). Societies construct in each case their world—but this entails that there is something possessing in itself this quality independent from any construction: that it is construct*ible* (in part, to be sure).

□ But we are in no position, from an ultimate point of view, to separate rigorously and absolutely disentangle that which, in these constructions, originates in the constructing subject—in this case, society—and that which appertains to the world in itself, to what there is. Our effort to achieve such a separation is certainly neither sterile nor meaningless, on the contrary; but it is bound to be interminable.

We can show this in a more precise way in the case of the social institution of time.

Society creates itself—institutes itself—along two intertwined dimensions: the ensemblistic-identitary (ensidic) dimension, and the properly imaginary, or poietic, dimension. It creates current, ordinary logic and arithmetic and objects with stable attributes and permanent relations: this is the identitary (ensidic) dimension of the institution, and of all the significations it embodies. This, society could not do without leaning on the first natural stratum, i.e., on the immediately accessible layer of the world, as given to humans by dint of their animal constitution. To put it bluntly: the existence of societies proves that there is in the world in itself—at least, in the first natural stratum—something corresponding to arithmetic and geometry. Or: causality is certainly an *a priori* category, not inducible from the phenomena. But causality would be useless if it were not, repeatedly, confirmed through the possibility of its application. The same is true concerning identitary social time, or calendar time: identitary social time is created leaning on cosmic time, that is, on the fact that equivalent recurrences (equivalent "sufficiently for use and need," as Aristotle would say) do exist among the data of the first natural stratum: they have to be singled out—but they *can* be singled out. So much, but also no more. For social ensidic time to be created and organized in detail, it is necessary that the first natural stratum exhibit what can be constructed as equivalent recurrences. Nothing is thereby said as to the temporality or otherwise of other layers of the world (neither, or course, as to their nature). The same must be said about the fundamental characteristics of usual time, e.g., local irreversibility, intransportability, etc.: they equally lean on aspects of the ensidic dimension of cosmic time. Whatever the riddles of the concept of irreversibility in physics, for instance, the fact remains that all the King's horses and all the King's men could not put a broken egg together again.

But on these beams, so to speak, a proper social time is, in each case, erected: the imaginary dimension of social time frequently overrules the above-mentioned characteristics. There is for instance no simple and absolute irreversibility for many, if not most, beliefs and religions (which posit a cyclical time); neither is there, necessarily, intransportability of the segments of time (or of time-bound processes) for many of them (shamanism, magic, etc.). The question about time and its characteristics beyond the data of the first natural stratum and beyond the beliefs of the tribe first emerges with the creation of philosophy and rational/scientific thought.

Imaginary time would be the most important theme to treat—and I cannot treat it here. Suffice it to say that it is, in each case, consubstantial with the most decisive aspects of the overall institution of society and its imaginary significations. And, as for all nuclear social imaginary significations, its content is essentially independent of any substantive leaning on the first natural stratum: it is a pure creation of the society considered (compare, for instance, Christian and Hinduistic or Buddhistic time). Taking this into account in its most general implications, our philosophical question about the world can be given a sharp formulation. In relation to identitary time, as to the whole edifice of identitary objects and relations erected by society, we ask: how must the world be, in itself, in order that this edifice can be erected. The only possible answer is: the world must contain the (otherwise mysterious) equivalent of an identitary dimension. For example, cows and bulls beget, and can only beget, calves and heifers in functional social life—irrespective of whatever bulls and cows may mean in religion or in the tribe's representations; stars return periodically and perpetually, whether they are gods, God-created luminaries, or heaps of hydrogen and helium.

In relation to imaginary time, as well as to the whole edifice of imaginary significations erected by each society, we ask: how must the world be, in itself, in order that this amazing and unlimited variety of imaginary edifices can be erected. The only possible answer is: the world must be tolerant and indifferent as between all these creations. It must make room for them, and for all of them, and not prevent, favor, or impose any among them over and against the others. In short: the world must be void of meaning. It is only because there is no signification intrinsic to the world that humans had, and were able, to endow it with this extraordinary variety of strongly heterogeneous meanings. It is because there is no voice thundering from behind the clouds, and no language of Being, that history has been possible. (Of course religions, especially revealed religions, assert the contrary. But there are just too many of them.) And the prevalence, today, of the Western imaginary signification of the unlimited expansion of pseudo-rational pseudo-mastery is made possible by the ubiquity of the identitary dimension of the world, on which its practical achievements lean—and which, as such, is meaningless.

IV

Social identitary time—and therefore, social time *tout court*—presupposes identitary time in the first natural stratum or an identitary dimension of time (as of the rest) of what there is in general. It is with this identitary dimension of the world that physics deals to begin with. I must limit myself to a few brief remarks on an immense, immensely difficult subject, inextricably linked with mathematical and physical technicalities.

There is no question that the specter of the spatialization of time haunts the whole of physics since, at least, the times of Lagrange ("physics is a geometry in four dimensions"). Einstein himself firmly believed that time is a subjective illusion (whatever that may mean). In mathematical physics, time appears centrally as the fourth dimension of a four-dimensional manifold. It is not easy to see *why* it is distinct from the other three dimensions, nor *what* distinguishes it from them.

Usually, irreversibility of time (or rather, of processes in time) is brought in in order to supply the proper character of time: movements in space are reversible, processes in time are not. But this is unsatisfactory from many points of view. First of all, it is not at all certain that all movements in space are reversible. Wherever there is a very strong gravitational gradient, for instance (as in the neighborhood of a black hole, say), spatial movement can take place only in some privileged directions (except for quantum effects). And, if we take the Universe as a whole and the prevalent cosmological conceptions explaining the observed red shift of the light emitted by distant galaxies, there are directions in space which can not be reversed: no cluster of galaxies could move "inwards" during a phase of expansion of the Universe, nor "outwards" during a phase of contraction. Secondly, as precisely the previous example hints at, irreversibility becomes a riddle on the cosmological scale. I cannot resist the temptation to quote a news item illustrating this in a somewhat amusing way (*New York Times,* Jan. 21, 1987):

> Of all phenomena that affect the human condition none has perplexed scientists more than the forward march of time, its link to the seemingly relentless tendency toward disorder known as entropy, and to the expansion of the universe.
>
> Some of the world's leading theorists have speculated that, if the current

> expansion reverses itself and the universe begins to contract, the arrow of time will change direction. People—if there are any—would live from the grave to the cradle and would "remember" what is to happen tomorrow. Some theorists have suggested that those living in such a universe would not be aware that time was running backward, because their perception of time would be reversed. But they would live in a universe whose future, in every detail, is predetermined. Scientists have also suggested that our universe might have a twin, formed of antimatter, in which time runs backward.
>
> Stephen W. Hawking of Cambridge University in England, a prominent proponent of the view that time would run backward in a shrinking universe, announced recently that he had changed his mind. Recent research has led him to conclude that time would still march forward, even if the universe began to contract, he told a conference in Chicago of astrophysicists.

With all due respect to the extraordinary mind of Stephen Hawking, there is a bit of consolation for the philosopher to see the prominent physicists of today caught in the tangles of time—and, one must add, exhibiting a modicum of *naïveté*. One must also deplore the waning of classical studies. Surely, seventy years ago, a Hermann Weyl or a Werner Heisenberg would not have missed the opportunity, in this context, to mention that "time running backwards" is the central theme of the famous myth in Plato's *Politicus* (a title wrongly rendered in the standard English usage as "Statesman").

Third, as is well known, attempts to deduce irreversibility from first physical principles (or even, to make it compatible with them), which started with Boltzmann a century ago, have remained unsatisfactory.

Fourth, and the most important, physical irreversibility is of course *locally* an undisputable fact. But it is a *partial* fact, which does not, by far, exhaust the data. More precisely, physical irreversibility is interpreted as increasing entropy, that is: homogenization and disorganization (which, let it be said in passing, entails the paradox that, were the tendency toward increase of entropy to prevail fully and the Universe to become, as it should, a photon gas, time would cease to have any physical meaning). But entropy is not all: living species emerge, babies are born and grow, painters put together masterpieces. All this does not "violate the second law of thermodynamics"—it just is beyond its scope. Forms are not only destroyed, they are also created, and one cannot reach an understanding of time, nor, I think, elucidate its "arrow" and irreversibility without taking into account both facts: creation and destruction of forms.

I have mentioned the specter of spatialization haunting physics. In fact, the question goes deeper. The spatialization of time in physics is but a consequence of the fact that physics, mathematical physics, treats everything within the ensemblistic-identitary (ensidic) framework, including space itself. This is a result of its dependence on mathematics (at least, mathematics up to now), which is the endless elaboration of the possibilities of the ensemblistic-identitary. Let me note in passing that this was also the error of Bergson, who criticized the physicist's conception of time as spatialized, and identified space with the quantifiable. This is only true for abstract—i.e., mathematical, ensemblistic-identitary—"space." Insofar as time is treated by physics as just the fourth dimension of a four-dimensional geometrical space, this is true. But nothing ensures that actual space (the space we live in as well as the space of the world in itself) is reducible to abstract, mathematical space (and is, thus, susceptible of quantification pure and simple). This is not the place to elaborate this question. Henceforth, my references to space as distinct from time have to be understood as relative to abstract, mathematical space.

We cannot reach the kernel of the question of time—be it subjective, objective, or overarching time—unless we start from the idea of the emergence of Otherness, that is, from alteration (*alloiōsis*), as creation/destruction of forms, considered as a fundamental determination of being as such, that is, in itself.

This forces us to distinguish strictly between difference and otherness. Thirty-four is different from 43, a circle and an ellipse are different. The *Iliad* and *The Castle* are not different—they are other. A horde of baboons and a human society are other. Human society, for instance, exists only as the emergence of a new form (*eidos*) and embodies such a form. We will say that two objects are different if there is a set of determinate transformations ("laws") allowing the deduction or production of this from that. If there is no such set of determinate transformations, the objects are other. The emergence of the other is the only way to give a more than verbal meaning to the idea of newness, or the new as such. The new is not the unforeseeable, unpredictable, nor the undetermined. Something can be unpredictable (e.g., the next number in a roulette) and still be the trivial repetition of a form; or be undetermined, and again, a sheer repetition of a given form (e.g., quantum phenomena). Something is new when it is the position of a form neither producible nor deducible from other forms. Something being new means therefore: something is the position of new determinations, of new laws. This is the meaning of form—of *eidos*.

The new *eidos,* the new form, is created *ex nihilo* as such. It is not, *qua* form, *qua eidos,* producible or deducible from what "was there." This does not mean that it is created *in nihilo* or *cum nihilo.* So, for instance, humans create the world of meaning and signification, or institution, upon certain conditions, viz., that they are already living beings, that there is no constantly and bodily present God to tell them what is the meaning of the world and of their life, etc. But there is no way we can derive either this level of being—the social-historical—or its particular contents in each case from these conditions. The Greek *polis* is created under certain conditions and "with" certain means, in a definite environment, with given human beings, a tremendous past embodied *inter alia* in Greek mythology and language, and so on, endlessly. But it is not caused or determined by these. The existing, or part of it, conditions the new form; it does not cause or determine it.

The fact of creation as such has nothing to do with the quarrel about determinism. It only contradicts the paradoxical, if not absurd, idea of a homogeneous universal determinism that could reduce levels or strata of being (and their corresponding laws) to a single ultimate and elementary level. This would entail *inter alia* the interesting metaphysical-theological conclusion that it was strictly necessary for the universe to reach a self-knowledge (by means of physical theory). Creation entails only that the determinations over what there is are never closed in a manner forbidding the emergence of other determinations.

This allows us to offer a characterization of time in its distinction from abstract, mathematical space (and space-time). We can abstract, in thought, from that which is different, and think of pure difference as such. This is possible—and the result of this abstractive operation is pure, abstract space. In this space every point is different from any other point without any intrinsic characteristic—just by virtue of something external to it, that is its position "in" space. Two strictly identical cubes are different if and only if they are in different places in space. Abstract space is this miracle, this fantastic possibility of the difference of the identical. Points, equal segments, figures, or solids can be distinguished without any "proper" difference—because they differ as to location, as to their position in space.

Difference is infinitely productive: e.g., it underlies and makes possible the whole of mathematics. In mathematics we proceed by attaching characteristics to sets of "indifferent" elements, then making these same characteristics "indifferent" on another level, and so on. "Production" means building up from given elements and according to given laws. We can think of an infinite manifold of "identical" elements along one "dimension" or along any number of "dimensions"—and we have a "space-like" receptacle. We can fill this receptacle with objects produced as different—i.e., reducible to each other and all to some elementary objects, according to determinate rules and laws: we then have a pseudo-physical, immobile universe. We can then add to it a supplementary dimension, call it time, and endow it with some peculiar properties, distinguishing it from the other dimensions of the pseudo-physical manifold. Such properties can be, e.g.:

(a) productions are irreversible: that is, the inversion of the total order structure imposed on this dimension is impossible or meaningless;
(b) there are some singled-out properties of the elements and the constructions which are invariant along this dimension—i.e., properties that are "conserved" along this "space-like" time (e.g., "matter-energy," and, now, some other, more exotic ones, in quantum physics);
(c) some subsets of "elements" and "productions," called processes, are not transportable along this dimension.

We then have a four- (or n-) dimensional manifold, constructed from the identical and the different (that is, the identical repeated), and which we can think of and elaborate abstracting from any concrete content of it. (Things become more complicated in general relativity, where the measure of "time" depends on the total "spatio-temporal" structure of the Universe, which depends, in turn, on its matter-energy "content." But then, of course, we run into the cosmological riddles, some of which were hinted at above.)

But in the case of otherness we cannot abstract from that which is, in each case, other; we cannot think of pure otherness as such. Otherness does indeed appear also in space—but there is no pure, abstract space for otherness. Otherness is always the otherness of something in respect to another something (*ti* and *allo ti*—*etwas anderes,* not *etwas verschiedenes*). We do experience otherness the moment we fall in love (or discover that we are already in love), or with any sudden change of mood, or in the emergence of another idea, or when we read *The Castle* after *Madame Bovary,* or look at photographs of the Parthenon and the Cathedral of Reims, or even when we look at a rock, and suddenly see a worm moving on the rock.

In this case, thus, we do not have a pure receptacle which can or cannot be filled with indifferent elements. The dimension along which otherness is, is, in each case, consubstantial and coemergent with that which emerges as other in respect to something. It is inseparable from it—from the forms or events which make the otherness be, and which make be, in each case, another otherness. The differences in the positions of Mars and Venus relative to the Earth are comparable (and therefore measurable in identitary space-time). The otherness separating *Gaspard de la Nuit* from the *Rasumowsky Quartets* and the latter from *The Art of the Fugue* is not comparable—and the chronological distance between these works (measured in identitary, calendar time) gives us only external benchmarks. Otherness is irreducible, indeducible, and not producible.

Insofar as the form emerging in each case is other, it brings with it—is consubstantial with—its own time. There is another time for each category or class of otherness. And there is always a question of a proper time for each instance or realization of the new form—even if it be unique. The time of the cell is certainly not the time of the organism as a whole; but also, the time of Flaubert's *Education Sentimentale* is not the same as the time of Beckett's *Endgame*. The encasing, nesting, interlocking of these times among themselves is a huge subject in its own right, which cannot be dealt with here. As emergence of the otherness—of that which cannot be produced or deduced from what is there—being is creation: creation of itself, and creation of time as the time of otherness and of being. And creation entails destruction—if only because another form alters the total form of what was there.

Difference and abstract space are solidary; but they are extrinsic to that which is, in each case, different: e.g., two points. Thus, we can think of abstract space, precisely abstracting from any particular content: mathematics. Otherness and time are solidary. But otherness and time are not extrinsic to that which is, in each case, other. We cannot think of pure otherness as such. An empty space is both a legitimate mathematical concept and a possibility of our intuition. An empty time is nothing—or it is just an additional "space-like" dimension, of which, considered as such, we cannot have an intuition and which simply we cannot think. I would add: irrespective of any possibility or impossibility of our intuition, an empty time cannot be.

Time is being insofar as being is otherness, creation, and destruction. Abstract space is being insofar as being is determinacy, identity, and difference.

A digression here is necessary. I have been talking about abstract space, and warned against the mistaken identification (Bergson) of abstract space with space *tout court*. What Bergson calls space is only true about mathematical space (and the space of mathematical physics), and in fact concerns the ensemblistic-identitary dimension of space. But such an identitary (ensidic) dimension is inherent to whatever there is—even to time, and this allows us (societies) to construct a public, identitary time (calendar time). The usual public time, as well as the usual public space, is constructed by society and endowed with definite ensidic characteristics (homogeneity, repetition, difference of the identical, etc.), leaning obviously on ensidic characteristics of what there is—beyond which there is certainly a vaster multiplicity about which, to begin with, we know nothing. But also, abstract space does not by far exhaust what we have to think of as space. Nothing authorizes us to treat space as through and through identitary. I am not only talking about the fact that actual space is never purely ensidic for a subject (animal, human, society, etc.), never reducible to homogeneity, repetition, etc., but always qualitatively organized and articulated by and for the subject (this is Heidegger's *In-der-Welt-sein*). I am talking mainly about the deployment of being as deployment of a heterogeneous multiplicity of coexistent alterities. Even the consideration of time as such brings us to this idea, since we have to admit the coexistence (and interlocking, mutual encasing, etc.) of a multiplicity of proper times. We have therefore to think of space as containing not only an ensidic, but also an imaginary or poietic dimension. Insofar as it entails the "simultaneous" deployment of forms which are other, insofar as it allows an "instant cross-section" of whatever there is as other, insofar as there is "synchronous multiplicity" of other forms, actual space, in the full sense of the word, goes beyond abstract space and beyond simply ensidic organization.

It would therefore be wrong simply to equate space (full space, actual space, as distinct from abstract space) with identity and difference, repetition, determinacy—in brief, with the ensemblistic-identitary (ensidic), and time with alteration, creation/destruction only. There is poietic space, space unfolding with and through the emergence of forms. And there is identitary time, ensidic time embedded in poietic or imaginary time. And it is the limit of this identitary time we vainly attempt to reach, when we try to think of the difference between the state *S* and the state *S'* of a pure photon gas. Even in this case, there would certainly be a difference, and this difference would be describable by and for an ultra-fine and ultra-powerful observer—whose appearance, however, would immediately destroy the state of the Universe as a pure photon gas and who, in addition, would be, through his subjective observations and acts, the only source of meaning for a before and an after attached to the states of the gas.

Is there, then, a possibility for an essential distinction between time and space—beyond the lived evidence of this difference, beyond the objectivistic reduction of time to abstract space and beyond the positivistic avoidance of the question? I think that there is, and it is grounded in their distinct relation to alterity and alteration.

I say: the emergence of forms is the ultimate character of time; the before and the after is given by the scansion of creation and destruction. Along this line we can, in a sense, elucidate irreversibility. In the indifferent, ensidic dimension of time—in the measurable but reversible repetition of the identical as the successive—forms emerge, or forms are destroyed (*not:* thermodynamically disorganized!). The direction along which disorganization of the ensidic (entropy) increases *and* forms emerge and are destroyed *qua* forms, gives us an arrow of time. (Forms *qua* forms are not necessarily destroyed by entropy. There is no possible meaning in the sentence "The Roman Empire collapsed because of the second law of thermodynamics.") Could we reverse this arrow? If we restrict ourselves to the identitary or ensidic dimension, such a reversal is only immeasurably improbable. But if we take into account forms, the idea of a reversal becomes meaningless. There is a finite (though vanishingly small) probability for the drop of ink diluted in a glass of water to condense again spontaneously in the exact place where it was dropped to begin with. There is no meaning in the idea that Proust could have written *La Recherche* before *Jean Santeuil:* or that Athens could have started with Demosthenes and proceeded, through Pericles, to Solon and his predecessors.

This is not because the after was caused by the before. In the most important cases, we cannot speak of causation; and, at the elementary level, the action of causality is reversible (this is the root of the difficulties in the thermodynamic "deduction" of irreversibility). It is because the before (the relevant, in each case, before) conditions the after in a nonreversible way. (The trivial but fundamental distinction between causes and conditions, or that between simply necessary and necessary *and* sufficient conditions, is surprisingly often forgotten in this type of discussion.) Forms as forms are not caused by something—but they emerge given certain (in fact, innumerable) conditions. The conditions allow the emergence of the form—but the converse is meaningless. Thus, the reversal of the arrow of time is extremely improbable from the abstract, ensidic point of view—and simply absurd, when the emergence of forms is taken into account. Not only can we not conceive the Greek *polis* without the Greek mythology; the *polis* was, in itself, impossible without this mythology (which far preceded it). But the mythology did not cause the *polis*—it was not the necessary and *sufficient* condition for it (even if supplemented by any number of other conditions); neither can we, or anybody, derive the one from the other, in either sense.

What is then the distinction between time and space? I said before that usual (thermodynamic) irreversibility does not suffice to establish this distinction. We speak about time as the emergence of forms, an emergence conditioned in each case by the (or some of the) already existing forms. But by the same token, forms emerge also—though in another sense—in space (not necessarily "physical" space) and can only be by deploying a space. We can even say, and it is obvious, that the emergence of a new form is conditioned by the (or some of the) forms surrounding it. Any here is conditioned by the elsewhere.

All the same, the distinction can be made.

The time perspective is effectively complete. It contains and entails space. In time forms emerge, are created; but a form is an organized multiplicity, thus its emergence brings into being a simultaneous coexistence (of the constituents of the form). The converse is not true. The space perspective is essentially deficient. The being of a form, considered as such, does not refer to or relate with any succession, any past/present/future; neither is it in need of time to deploy itself. (In a strange reversal, typical of inherited thought, this fact has been considered from Plato onwards as "grounding" the "derived" character of time.)

We can express the same idea in yet another way.

If no new forms were to emerge we cannot say that space would cease to exist; not even that it would become abstract, ensidic space. We can conceive of a heterogeneous space, full of immutable forms, other and other, in which nothing happens. (A Platonic world of Ideas could be a model for such a space.) If a voyager were to go around this space and find successively other forms, each new to him, these would be *his* discoveries; nothing would still happen, except his (impossible) voyage through this space and the changes in *his* subjective states (scansions in his subjective time, without relation to the world he is visiting. This is more or less the journey of the Platonic soul in the supracelestial world.)

But we can say that, if emergence of alterity, creation/destruction of forms, were not there, there would be no time (except in the impossible purely ensidic sense explained above). Bringing this thought to its limit, we can say that no thing (nothing) would be there, since no form would ever have arisen.

In this sense, time is essentially linked to the emergence of alterity. Time is this emergence as such—whilst space is "only" its necessary concomitant. Time is creation and destruction—that means, time is being in its substantive determinations.

V

We have posited two fundamental categories to help us elucidate the question of time: difference and otherness. We now may bring together difference and otherness under multiplicity. Multiplicity formally entails unity; without unity, multiplicity would not be multiplicity—it would be an uninspectable, in itself dispersed and disconnected Infra-chaos. Unity, on the other hand, does not entail multiplicity. It just happens that there are many. It just happens that being is—and that it is not one. This we can only see and accept, we cannot elucidate it further.

What does it mean that being is—and is not one? Insofar as multiplicity in being exists as difference, being is one not only logically and nominally (being as abstract name for whatever there is), but effectively. Multiplicity as difference means that the plurality of particular beings is brought into one by the laws which produce, deduce, etc., beings from each other. Briefly and brutally speaking, qualities are reduced to quantities and different quantities give different (reducible) qualities. This is both Hegel and the dominant, reductionist trend in the positive sciences.

But insofar as multiplicity in being exists as otherness, or alterity, the unity of being is essentially fragmented. This is because, despite all the recent talk about the ontological difference, being and mode of being are not separable—and modes of being emerge, thereby altering being itself, and manifesting being as self-alteration. To be sure, emergence as such is distinct from that which, in each case, emerges—as presence is different from the present, and being from beings. But this is a scholastic, logical distinction. Being as self-alteration entails also alterity in the modes of emergence—and talk about emergence as such, abstracting from the mode of emergence which is in turn inseparable from that which is emerging, would be empty talk. Such has been Heidegger's talk about being, or about presence. Presence as such—the fact of presence—is certainly distinct from that which is present; but modes of presence are other, and there can be no thought of presence as such abstracting from the modes of presence. Not only can we not put under the same title—except verbally and vacuously, as Aristotle would say—*The Well-Tempered Clavier* and the Andromeda Nebula; we cannot think being as self-alteration and incessant to-be without considering the modes of this self-alteration and the modes of being they bring about.

Does being exist as otherness, alterity? Certainly; if it did not there would not be a being-subject (indefinitely many beings-subjects, and indefinitely many modes of being-a-subject), creating each time its own mode of being and its own world (and time), and, for instance, thinking and talking about being. Without otherness, there would not exist a question of being. Not only would there be nobody to ask the question, but, if the question were raised so to speak in the void, the answer would be simple: being would be a set, or a set of sets, and in this case being and mode of being coincide, as do possibility and actuality. Mathematically, *is* simply what *is possible,* and something is not if and only if it is impossible. Elements of a set are if and only if a set can be defined, in a consistent way, of which they are elements.

Multiplicity of being is an irreducible, primary datum. It is a given. But what is also given is that multiplicity exists as difference on the one hand, as otherness on the other hand. Insofar as difference is a dimension of being, there is identity, persistence, repetition. Insofar as otherness is a dimension of being, there is creation and destruction of forms. And indeed, otherness entails difference. A form cannot be said to be unless it is identical to itself (in the broadest sense of the term "identical"), and persists/repeats itself for a while—that is, in and through an identitary dimension along which it differs with itself only by being placed in a different (identitary) time. And this is but one aspect of the fact that no form can be without a minimal determinacy. That means that any form has necessarily an ensidic dimension—and therefore participates necessarily in the ensidic universe.

If then these are the characteristics of being, we find that they are the same as the ones we should attribute to time: the unfolding of otherness, the deployment of alterity, together with a dimension of identity/ difference (repetition). The latter alone we find in abstract (ensidic) space. We find both—difference and alterity—in actual space; but, for the reasons given above, actual space presupposes time. The fullness of being is given—that is, simply is—only in and through the emergence of otherness which is solidary with time.

With self-deployment in and through time, that is, with the emergence of otherness, we can understand that the unity and unicity of being are truly fragmented and stratified. This is particularly manifest with the emergence of being-for-itself (starting with the living being) which entails the creation of other modes of being (objectively) and of other, self-closed worlds (subjectively), with, in each case, their own time. The being-for-itself unfolds also, *qua* being, in space and time. But the being-for-itself creates time and space and being for-itself, and thus it fragments being, space, and time. And we cannot consider one temporality as the only originary or authentic one (such as the *ursprüngliche Zeitlichkeit des Seins-zum-Tode* of Heidegger's *Dasein*—which is of course typically a subjective temporality, exactly as its *in-der-Welt-Sein* is a subjective mode of a being in a *Lebenswelt* which is socially-historically created without *Dasein* or, for that matter, Heidegger, being aware of this fact) because we know and cannot pretend that we do not know that there is time for the living being and that there is cosmic time and that there is nothing derived or inauthentic about them.

As a result there arises, both for us and in itself, the question of the unity and unicity of being and time above and beyond their indefinite and unforgettable fragmentation and stratification. Insofar as the ensidic dimension alone goes, we could talk of a unity of being. But this unity is of course only partial, and, for the most part, inessential. (In both a Beethoven sonata and in a star we can distinguish enumerable elements. So what?) Thus, the overarching question of overarching time and being has to remain a question for the time being, and probably for all times.

PART II

Temporal Frames of Inquiry

A Slip in Time Saves Nine: Prestigious Origins Again

JONATHAN Z. SMITH

EACH YEAR, ON September 15, the *Dies Natalis* of the Temple of the Capitoline Jupiter, a Roman state official would hammer a nail (the *Clavus Annalis*) into the temple wall. This act marked time, in the passive sense of the term; it served to record each passing year like a prisoner's scratching on the wall of a cell. The ritual marked time in a more active sense as well. The nailing was an act of human inscription, interrupting the flow of time, fixing it for an instant, marking it as a human phenomenon. Because of the ritual, time is no longer perceivable solely as some external force (whether understood as "natural" or "divine") requiring submission; it is now subject, as well, to human deed and intention. The nail, after all, did more than mark time; it memorialized an act of human construction.[1]

Religion, as a part of its labor of cultural creation, works constantly with time in highly complex and various fashions. By contrast the study of religion, as a part of the academy's labor of cultural creation, has tended to represent the work of religion with respect to time in a simple geometry.[2] You know the drill: circles and lines, "that's all the facts when you come to brass tacks." This geometry has often been extended into a highly valenced taxonomy: the familiar dualisms of "myth" and "history," of "archaic" and "modern," of "closed" and "open," and other variations of the same.

The development of this duality occurred at a quite specific moment within the academy. In the mid–nineteenth century, European thought underwent a revolution with respect to time comparable to the revolution with respect to space that resulted from the Age of Exploration (or better, the Age of Reconnaissance) in the fifteenth through seventeenth centuries. Both revolutions produced a series of cognitive shocks to European thought and sensibility. The history of the prior, spatial revolution—above all the encounter with "other" cultures—and its effects on the human sciences has been insistently told. The history of the temporal revolution has yet to be fully written.

The initial, and better-known, moment in the temporal revolution was the series of geological and palaeontological discoveries and arguments that culminated in the notion of the antiquity of humankind as a natural species with its concomitant shock both to biblical and philosophical traditions. The second moment, which followed in quick succession, was in many respects the converse of the first. It was the discovery of the antiquity of humankind as a recognizable and familiar cultural species. I refer here to the extraordinary series of successes in the decipherment of the languages of the ancient Near East: first, Egyptian hieroglyphs (1822–24) and Akkadian (1849–57); then, more recently, Sumerian (1905–23), Ugaritic (1930–31), and Hittite (1915–42). The first moment discovered incalculable temporal distance and cultural opacity, the silence of bones and stone artifacts encapsulated in the word "prehistory." (Since the introduction of this term into English by Daniel Wilson in 1851, we have become so accustomed to it that we fail to flinch at its shock.) In stark contrast, the second moment revealed supremely articulate cultures: human beings who could speak directly to us, who could tell their story in their own words through the technology of inscription. The decipherment of these writing systems was a triumph of scholarship; it was also a severe cognitive shock, one with which, even today, within the academy, we have by no means made our peace. For in revealing people very much "like us" it forced us to alter our understanding of our cultural roots. Israel and Greece were not, as had been hitherto thought, the foundations of Western civilization. They were later, secondary, perhaps even derivative cultures. "High," "oriental" cultures had preexisted them by millennia.

At the forefront of this second moment in the temporal revolution was a group of German scholars who came to be known as the Pan-Babylonian School. Among their many claims, most of which have been rejected by more contemporary scholarship, was one that, curiously, has survived criticism: their insistence that the essential *Weltanschauung* (a term largely popularized by the school) of the ancient Near East as revealed in the cuneiform texts was based on an uncommon faith in the cyclical order of the heavenly bodies and an imperative to harmonize earthly affairs with this order. "As above, so below." When expressed by the ancient texts in astronomical terms, this view was named by the Pan-Babylonian School the "Babylonian" pattern. Brought down to earth and expressed in the language of the alternation of the seasons rather than the movements of the planets and stars, it was called the "Canaanite" pattern. If the cycle was stretched into an ellipse and then a line, and focused on human rather than "cosmic" or "natural" activities, it was termed the "Israelite" pattern. The first two patterns, the school maintained, were relentlessly cyclical: in the Babylonian pattern, the cycle was expressed as an alternation between chaos and creation, in the Canaanite pattern as an alternation between winter and summer. The Israelite pattern, by contrast, was linear, moving from a singular beginning to an equally singular end. The Pan-Babylonians identified the cyclical patterns as being essentially "mythic" and the linear pattern as essentially "historic." To the degree that their project has been taken as self-evident by scholars in religion ever since, as it has, the world-view of the academic study of religion remains relentlessly Pan-Babylonian.[3]

The success of this model is astounding as it in no way accords with the data from which it was first generated or with the bulk of the data to which it has been subsequently applied. A notable example is the Gilgamesh Epic, first edited by one of the leading members of the Pan-Babylonian School (P. Jensen), and interpreted (fantastically) by the school as a solar hero's astral journey through the circle of the zodiac. The entire reason for introducing the hitherto independent flood story into the eleventh tablet of the late Akkadian version of the narrative is to make the point, fundamental to this recension of the epic, that the flood was an irrepeatable event and that, therefore, the immortalizing of Utanapishtim would not be repeated in the case of Gilgamesh (XI, 197–98). Even a divine king could not escape death. With respect to the whole question of Near Eastern literature, one must agree with the careful conclusion of H. Gese: "No matter how often it is repeated, a reference to the entanglement of the ancient Near East in an a-historical nature-myth of the eternal cycle of all events is meaningless in light of the fact that such a mythology simply does not exist in the historiographic documents of the ancient, i.e., pre-Persian Near East."[4] Nor, for that matter, does such mythology exist in any other genre of Near Eastern literature. Indeed, I am more confident of the presence of *Urzeit/Endzeit* correlations in late Israelitic prophecy (e.g., Isaiah 51:9–11) than in any of the mythic texts of Mediterranean antiquity.

Why, then, has the putative distinction had such success? Surely not because it enables the interpretative enterprise, but rather because it fulfills an ideological role. The prime use of the construct has been to protect the Bible from the enterprise of comparison, whether genealogical or ontological, and then, by extension, to protect Judaism and especially Christianity from the effects of the comparison of religions, to insulate "us" from "them." Thus the familiar formulations: neither Israel nor Christianity "borrowed" from its environment; they are "historic" rather than "mythic" religions; "God Acts in History."

Decoded, these various characterizations reduce to a set of insistent dualities that make impossible the interpretation and valuing of the bulk of the religious expressions and activities of humankind. We, the scholars proclaim, stand under the sign of the "unique," they stand under the sign of the "repetitive." We are "original," they are "mimetic." We participate in transcendent "culture," they remain enmeshed in "nature." Or, to state the opposition in its bluntest form, we are human, they are not. For, as long as we identify recognizable humanity with historical consciousness and identify critical thought with openness to change—as we do—the usual treatment of the religions of other folk is, designedly, inhumane. It is the scholar's gaze that has imposed the claimed circularity and ahistorical character on myth. By doing so, we have robbed the "other," whether thought of as "primitive" or "oriental," of indigenous capacity for thought, especially with respect to the hard work of cultural creation.

These various themes are brought together in an all-too-typical paragraph from an article by a distinguished American biblical scholar, Bruce Metzger, devoted to the methodology of comparison:

> It is generally acknowledged that the rites of the mysteries, which commemorate the dying and rising deity, represent the cyclical recurrence of the seasons. In other words, such myths are the expression of ancient nature-symbolism; the spirit of vegetation dies every year and rises every year. According to popular expectation, the world process will be infinitely repeated, being a circular movement leading nowhere. For the Christian, on the other hand, as heir to the Hebraic view of history, the time-process comprises a series of unique events, and the most significant of these events was the death and resurrection of Jesus Christ. Unlike the recurrent death and reanimation of the cultic deities symbolizing the cycle of nature, for the Christians, the importance of Jesus' work was related just to this "once-for-all" character of his death and resurrection.[5]

While there is much in this paragraph that is deserving of criticism, I shall dwell, at the outset, on only two items—one general, one specific—that recur throughout this sort of regnant religious scholarship.

First, to begin with the general item, it is long overdue that we set aside the notion of "nature mythology" that is at the heart of the mischievous distinction between "mythic" (i.e., cyclical) cults and "historical" religions. Such a notion is based on the old, inadequate idea of myth as bad science. Ironically, the very sort of "confusion" lifted up by Victorian scholars to classify "primitive" thought as "magical" or "prelogical" has been more often characteristic of modern "scientific" scholarship on religion: native statements of correlation have been taken for assertions of identity.

As we have come to learn, myths are not about nature. They are not best understood as "primitive" attempts to explain natural phenomena. Myths often think *with* natural objects; they are almost never about them. Their focus is not on the genealogy of things but on the topography of relationships. This has been put most bluntly by C. Lévi-Strauss: "The mistake of Mannhardt and the Naturalist School was to think that natural phenomena are *what* myths seek to explain, when they are rather the *medium through which* myths try to explain facts which are themselves not of a natural but a logical order."[6] In the case of the dying and rising deities considered by Metzger, the seasons may serve as an experimental medium for thinking about issues such as periodicity, regularity, transformation, and the like. This is, above all, an intellectual activity that need not be equated with agrarian concerns or fertility cults.[7]

Second, the specific category of "dying and rising" gods, like the general notion of "nature mythology," can no longer be sustained. In every case that has been proposed, it has become clear that the deities so named have died and are mourned by their cults; there is no account of their rising until late texts from the Christian era.[8] While this new understanding of the data raises questions as to the soteriological significance of dying (but not rising) gods as well as the reasons for their later reinterpretation,[9] it has immediate effect on the sort of position represented by the passage quoted. However they are now to be understood, those deities formerly termed "dying and rising" gods cannot be used as self-evident witnesses for the ubiquity of religious preoccupation with seasonal cyclicality. Indeed, more careful research has demonstrated that these gods were not identified with vegetation until late allegorical treatments, the latter authored by individuals who were not adherents of their cults.[10] And there is more. If these observations be accepted, the distinction proposed between such deities as "mythic" and the Christian myth as "historical" collapses. Whether we speak of an Attis, an Adonis, or what have you, their death occurred "once-for-all"; it was not repeated in their mysteries. That there are differences within and between the various

Mediterranean religions I have no doubt; what is equally certain is that this difference can no longer be expressed as a matter of circles and lines, by the old duality of myth and history.

If these two observations be accepted, much of the content of Metzger's paragraph quoted above collapses. But one interesting element remains, deserving of further reflection since it leads to a set of issues that far transcend the particular question of concern to a biblical scholar: "The world process will be infinitely repeated, being a circular movement leading nowhere." That is to say, meaning is identified as singularly teleological—conceived either as giving a directionality to or as a "breaking in" on a lineal temporal movement. What is being asserted is that repetition is inherently meaningless, or, to translate this into other terms, repetition is the realm of the habitual. Meaning is to be identified with the singular, or, in more explicitly religious terms, with the epiphantic.

The limiting case of such a style of discourse within the field of religious studies is most certainly the theoretical position adumbrated by Adolf Jensen and other members of the Frobenius School (e.g., W. F. Otto). Jensen claims that all truth, meaning, and value are to be located in what he terms a primal moment of ontic "seizure" (*Ergriffenheit*), a "revelation," a "direct cognition of the essence of living reality." Myth, he argues, "always begins with a condition antecedent to concretization." The first "concretization" is an "intuitive spontaneous experiencing," which Jensen terms "expression." Expression is, for Jensen, an essentially passive act of reception. All subsequent "formulations" and "concretizations" (i.e., myth, ritual, and the like) are active "cultural" reinterpretations of this primal experience and are termed by Jensen "applications"—the most negative word in his lexicon. Application leads over time to "mere survivals" of the original seizure. All application falls under the iron "law of degeneration," "pauperization," or "semantic depletion," and causes the original "spontaneity" to become a "fixed but no longer understood routine."[11]

Note well the argument. Extraordinary privilege and priority is given to naked experience. Jensen locates all meaning in a prelinguistic, highly individual, originary moment of "seizure" that is *sui generis*. All socialization of this moment, from linguistic expression to repetition, destroys spontaneity, impoverishes meaning, and is, ultimately, mere habit. That which has been held by many to distinguish humankind from the animal (language) here renders humankind indistinguishable from the animal (creatures of irrational habit). One does not have to go far to discern the governing premise of this sort of theory. It is the Protestant principle of individuation, of unmediated and spontaneous revelation run rampant. Nor does the extremity of the position preclude one's finding other theories of the same sort of cunningly transformed "law" and "gospel" oppositions.

With Jensen as a limiting case, one may turn to a variety of other significant theorists and discern modulations of the same argument about time. Take two figures often juxtaposed to each other in the matter of the distinction between sacred and profane—Eliade and Durkheim.

In Eliade, space and time are conceived as parallel with respect to the difference between the sacred and the profane.[12] In both cases, the profane is "homogenous" and lacks "reality." Because of this, it is the realm of the "meaningless." This is not to say that there is no differentiation in this realm. Within the profane, privileged places (e.g., one's birthplace) are recognized, as is the alternation between "the comparatively monotonous time of . . . work, and the time of celebrations and spectacle, in short 'festal time' "—this latter is, I take it, an effort by Eliade to incorporate and distinguish himself from Durkheim. But these relative differences (of what Eliade terms "pseudoreligion") are not to be confused with the absolute distinction between sacred and profane. Sacred space and time are utterly discontinuous with profane space and time. They are "transhuman," "wholly different" in structure and origin. The sacred and the profane are two parallel modes of being in the world; the one, a mode of "eternity, substance and of reality"; the other, a mode of "the temporal, of change and illusion."[13] Given this split, humankind has an "unquenchable ontological thirst" to "found" its existence in "real existence," in "objective reality," in that which is not "illusory" and which avoids the "paralysis" of "the never ceasing relativity of purely subjective experiences." This reality is "given," it is "autonomous," it "*s'impose à l'homme du dehors.*"[14]

Eliade has a rich vocabulary for describing this "imposition." At times he speaks of the "manifestation" of the sacred that "ontologically founds the world." This is the sacred, not, as before, as a mode of perceiving the world, but as an ontological power that, in a highly dramatic fashion, breaks into the profane and "displays itself." When he writes in this fashion, he employs a language of "irruptions" and "elementary manifestations," those fundamental revelations that Eliade describes with words built on the root *phaino:* hierophanies, kratophanies, epiphanies, and the like. The human stance in the face of such a "display" is silence.[15] Thus far, although the vocabulary is different, Eliade shares the fundamental ontology and epistemology of Jensen. But with Eliade, all is not passivity and dualism. It is a paradox that the sacred should manifest itself in the profane—or, to unmask the language, that there should be a kenotic incarnation. This "rupture of ontological levels" will ultimately result in the swallowing up of the one by the other, in the "abolition" of the "spheres that are profane."[16] While this seems to be a project wholly from the side of being, there is human activity as well. The experience of the sacred is "infinitely recoverable and infinitely repeatable." Repetition is creativity. "Sacred time" is not the time of "irruption" (of revelation), but rather the time of beginning as made accessible through ritual. It is possible for the "stupendous instant in which a reality was for the first time fully manifested" to be "reexperienced" in creativity with respect to the world. Myth and ritual, insofar as they are repetitive, permit this "recollection."[17]

If Jensen offers an essentially Protestant model, Eliade may be seen as attempting to rewrite it as a Catholic-Orthodox one—with an assist from a combination of Platonic and Indic metaphysics. The notion of the paradox of the kenotic incarnation is central, allowing Eliade to venture a sacramental interpretation of human action. Repetition as "same" may well be meaningless; but he insists on another possibility: repetition as a recovery of originality, expressed as either a naked (prelinguistic) experience or as the human symbolization of an initial act of creativity that has the capacity to "point" beyond itself. From this perspective, the distinction between Jensen and Eliade is blurred. For both, the cost of retaining the privilege of a theological ontology of power and being has been an impoverishment of the anthropological (often termed by Eliade the "historical").

Although Durkheim clearly supplies the anthropological dimension so notably despised in Jensen and relativized in Eliade, one must observe the essential lineaments of the best-known segment of his argument: that concerning the distinction between the sacred and the profane.[18] An essentially spatial distinction—the sacred is that which may not be touched by the profane—is converted into a temporal one when he inquires into its etiology. So absolute a distinction cannot come from observation of the "natural" world. The answer, you will recall, builds on Mauss's study of the seasonal variations of the Eskimo,[19] and distinguishes everyday time from festival time: "How could such experiences . . . fail to leave in him [the aborigine] the conviction that there really exist two heterogenous and mutually incomparable worlds? One is that where his daily life drags wearily along . . . [the other, where he is excited] to the point of frenzy. The first is the profane world, the second, that of sacred things." The price of the description of the "effervescent" festal experience of the sacred—

where communication is reduced to the ecstatic "piercing cry, a veritable yell, 'yrrsh! yrrsh! yrrsh!' "—is the reduction of the everyday time of profanity to an inhumane level, bereft of all of the characteristics that distinguish, for Durkheim, humankind from the realm of the animal. It is a time of social "dispersion," almost wholly given up to "economic activity," where life is "uniform, languishing and dull."

In one sense, this description bears an uncanny resemblance to the model presupposed by Jensen and, to a lesser degree, Eliade. From another vantage point, it is profoundly different. I have already mentioned Durkheim's high valuation of the social and the anthropological. The same could be said for his valuation of the processes of "representation" and for his nascent linguistic theory of the sacred, which resists all ontological reformations.[20] Despite these meliorating differences, in Durkheim, inarticulate affective experience is still what validates the sacred, what gives it its sense of difference over against the sameness of the profane. The only element that prevents this from collapsing into the Protestant model is the priority of the social over the individual.

What is lacking in all three of these modulations of essentially the same model is a sufficient appreciation of the mundane. Each exhibits a failure to recognize that, within the domain of ritual, "God lives in details" rather than in high moments of drama. Ritual stands in a parallel relationship to ordinariness, not in opposition to it. As I have argued elsewhere:

> Ritual relies for its power on the fact that it is concerned with quite ordinary activities placed within an extraordinary [i.e., marked off, focused, arbitrarily delimited] setting, that what it describes and displays is, in principle, possible for every occurrence of these acts. But it also relies for its power on the perceived fact that, in actuality, such possibility cannot be realized. . . . Ritual [thus] gains force where incongruity is perceived and thought about. . . . Ritual is a relationship between "nows"—the now of everyday life and the now of ritual place; the simultaneity, but not the coexistence of "here" and "there." . . . This absolute discrepancy invites thought but cannot be thought away. One is invited to think about the one "now" in terms of the other; but the one cannot become the other. Ritual precises ambiguities; it neither overcomes them nor relaxes them. [For this reason] ritual, concerned primarily with difference, is, necessarily, an affair of the relative.[21]

The time of ritual (or myth) is not an "other" time ("sacred," "originary," "pristine") but a simultaneously different time. As we are learning to recognize here, as elsewhere, between a postulation of "otherness" and a postulation of "difference" there is all the difference in the world. "Otherness" is an absolute category that by its very nature forbids comparison and hence, as in Jensen and Eliade, intellection. "Difference" is a relative category that invites, indeed requires, the intellectual activities of negotiation, classification, and comparison while, at the same time, resisting any easy and unilluminating language of the "identical" or the "same." As I have suggested above, it is this thought about relationships, it is this comparative endeavor, that supplies the dynamics of both myth and ritual.

In characterizing the more usual disciplinary construction of religious studies with respect to time, I have deliberately employed the term "Protestant," for it is in Protestantism, and in Protestantism alone, that one finds the systemic articulation of the model of the "circle and line" in curious combination with a strong affirmation of "breakthrough." The Protestant mythos stands under a double burden of precedent: the old Christian burden of "Israel" and the new Reformation burden of the Roman Catholic Church. Within the mythos, this burden is expressed in terms of the "same": "law" and "pharaisaic ritualism" in the case of Judaism; "superstition," "magic," and "pagan ritualism" in the case of Catholicism. Into this mythic construction were inserted two privileged moments of "irruption": that of the "pristine" church of "primitive Christianity" (i.e., Paul) and that of "restoration" by means of a "return" from "corruption" to this origin (i.e., the Reformation). Ironically, I know of no non-Christian myth that expresses the real possibility of the recovery of "pristine origins" as strongly as does this mythos represented by early Protestant historiography with its assertion of living in a "fifth," reformist age within church history "*in qua Deus ecclesiam iterum ad fontes revocavit.*"[22] The Protestant model, which has been thoughtlessly extended to interpret the rest of the religious expressions and activities of humankind is, in fact, the exception that requires explanation.

Both the cyclical and linear models of time employed by most students of religion under the influence of the Protestant myth reduce "difference" to a discourse of the "same." The cyclical model does so by insisting on a notion of repetition that is held to be either inherently meaningless (as in Metzger) or meaningful because it is wholly congruent with its exemplar, that is to say because it lacks difference (as in Eliade). The linear model does so by reducing history to a succession of meaningless events, identical one to the other unless interrupted by or directed toward a supramundane *telos*. By viewing the sacred as "other" and the profane as the "same," the bulk of studies in religion have eliminated the complex middle ground of thought about "difference" where myth and ritual live.

The Time of Telling and the Telling of Time in Written and Oral Cultures

JACK GOODY

The Time of Telling: The Written

IN THE LITERATURE OF the high cultures we know, narrative plays an enormous role.[1] The novel, the film, the play, the opera, nearly all have a basically narrative structure in which time is of the essence. So too outside the literary sphere, most of the material in newspapers is constructed in the same way, the telling of news (the day's news), the systematic account of what happened, where references to time are omnipresent.

In his study *The Rise of the Novel,* Ian Watt saw the development of this genre in the eighteenth century as connected with the radical changes in English society that promoted the growth of realism, of individuality, and of a new concern with time. No longer relying on traditional narratives, new plots were created that had to be "acted out by particular people in particular circumstances" rather than by "general human types" set in "timeless stories." The genre developed two features of special importance: characterization and the presentation of background. Both were intimately concerned with the particularities not only of names but of time and place. "The characters of a novel," writes Watt, "can only be individualised if they are set in a background of particularised time and place."[2] Since Locke's *Essay Concerning Human Understanding,* personal identity has been seen as an identity of consciousness through duration in time.

Watt had in mind first and foremost that prototypical English narrative of the period, Defoe's *Robinson Crusoe,* where time is not simply sequence or duration, the kind of time found in earlier narratives, but a dimension capable of precise measurement. Defoe establishes the time and place of events in a way that provides an illusory realism, almost taking the novel outside the realm of fiction altogether, making it not simply the story of . . . , but the history of. . . . In *Robinson Crusoe* the relationship between the telling of the tale and the telling of time exists both for the writer and for his creature. "The editor," writes Defoe in that deceptively casual way, "believes the thing to be a just history of fact; neither is there any appearance of fiction in it." History demands dates. "I was born in the year 1632, in the city of York, of a good family, tho' not of that country."

Marooned on a desert island, Crusoe became deeply concerned with time, and to control its passage, or his understanding of its passage, he needed pencil and paper. At first his only companions were a dog and two cats, but they served him well, especially the dog. "I wanted nothing that he could fetch me, nor any company that he could make up to me, I only wanted him to talk to me, but that would not do. As I observed before, I found pen, ink, and paper, and I husbanded them to the utmost, and I shall shew, that while my ink lasted, I kept things very exact, but after that was gone I could not."

Keeping things very exact meant above all telling the time:

> After I had been there about ten or twelve days, it came into my thoughts that I should lose my reckoning of time for want of books and pen and ink, and should even forget the Sabbath days from the working days; but to prevent this I cut it with a knife on a large post, in capital letters, and making it into a great cross I set it up on the shore where I first landed, viz. "I came ashore here on the 30th of Sept. 1659." Upon the sides of the square post I cut every day a notch with my knife, and every seventh notch was as long again as the rest, and every first day of the month was as long again as that long one, and thus I kept my kalander, or weekly, monthly, and yearly reckoning of time.

That is to say, individual, personal, private markings have to operate within a social, public system for the graphic representation of time, by means of a calendar, which is at once a secular aid and a ritual, liturgical program. The measurement, the reckoning, even the conception of time in graphic terms involves giving that dimension a visual and spatial counterpart. Oral discourse takes place in time, written discourse in space and time; the latter is seen, the former unseen.

Keeping things very exact also involved another fascinating use not of graphics but of writing itself, namely laying out in a list the good and evil aspects of his present situation, exactly like debts and credits, as he remarks, in order better to understand the situation so that he could apply himself to "accommodate" his way of living.

Crusoe did obtain some books from the wreck, but above all he got pen, ink, and paper and began to keep a record of time. But he used his literate skills for yet another purpose: to draw up "the state of my affairs in writing." This he does by making a list or table, setting Evil on one side and Good on the other, just like "debtor and creditor." Lists and tables of a less semantically complex kind were the currency of many early written civilizations;[3] indeed, they are characteristic literate achievements, comparing synchronically what speech lets one examine only diachronically. The result was remarkable, for it enabled him to weigh up his condition, externalize it on paper, and then get on with his way of living. The whole passage is worth quoting as a dramatic illustration of the role of writing at a level that is at once both cognitive and practical.

> I now began to consider seriously my condition, and the circumstances I was reduced to, and I drew up the state of my affairs in writing, not so much to leave them to any that were to come after me, for I was likely to have but few heirs, as to deliver my thoughts from daily pouring upon them, and afflicting my mind; and as my reason began now to master my despondency, I began to comfort my self as well as I could, and to set the good against the evil, and I might have something to distinguish my case from worse, that I stated it very impartially, like debtor and creditor, the comforts I enjoyed against the miseries I suffered, thus:

I am cast upon a horrible desolate island, void of all hope of recovery.	But I am alive, and not drowned as all my ship's company was.
I am singled out and separated, as it were, from all the world to be miserable.	But I am singled out too from all the ship's crew to be spared from death; and He that miraculously saved me from death, can deliver me from this condition.
I am divided from mankind, a solitaire, one banished from humane society.	But I am not starved and perishing on a barren place, affording no sustenance.
I have no clothes to cover me.	But I am in a hot climate, where if I had clothes I could hardly wear them.
I am without any defence or means to resist any violence of man or beast.	But I am cast on an island, where I see no wild beasts to hurt me, as I saw on the coast of Africa; and what if I had been shipwrecked there?

I have no soul to speak to, or relieve me.	But God wonderfully sent the ship near enough to the shore, that I have gotten out so many necessary things as will either supply my wants, or enable me to supply my self even as long as I live.

Upon the whole here was an undoubted testimony, that there was scarce any condition in the world so miserable, but there was something negative or something positive to be thankful for in it; and let this stand as a direction from the experience of the most miserable of all conditions in this world, that we may always find in it something to comfort our selves from, and to set in the description of good and evil, on the credit side of the account.

Having now brought my mind a little to relish my condition, and given over looking out to sea to see if I could spy a ship; I say, giving over these things, I began to apply my self to accommodate my way of living, and to make things as easy to me as I could.

Just as writing enabled Crusoe to conquer (that is, organize) time, so too it enabled him to conquer his difficulties and organize his thoughts.

Defoe's concern with time in constructing narrative was by no means unique. The timing of events dominates Laurence Sterne's *Tristram Shandy* (1760) in a more personal but equally exact manner, beginning as it does with the unfortunate incident of his mother's interrupting his father at the moment of their monthly coitus: "'Pray, my Dear,' quoth my mother, 'have you not forgot to wind up the clock?'" His father was "one of the most regular men in everything he did," so that his son could declare with certainty, "I was begot in the night, betwixt the first Sunday and the first Monday in the month of March, in the year of our Lord one thousand seven hundred and eighteen."

The fine exactitude of timekeeping of course paralleled that of bookkeeping, the meticulous keeping of books. It was scarcely available to writers or desired by their readers until the grandfather clock of Shandy's father and more public varieties in the town hall became part of everyday consciousness, controlled by that tick-tock mechanism, the verge-and-foliot, the vertical escapement. Minimal but immensely important, that was the achievement of the West, and one of its few early technical exports to the East, apart from the universally acceptable firearms with which it has long continued to dominate the world. The verge-and-foliot mechanism that gave birth to the modern clock was not of course the first means of establishing, that is, measuring, hours (minutes, seconds); the water clock and the candle preceded it. But it initiated the domestication, the personalization, of this system, leading to that quasi-universal aspect of dress in the modern world, the wristwatch, which has become an intrinsic part of our apparel. And this particularization of time occurred in both everyday and literary contexts.

The Time of Telling: The Spoken

Beyond the realm of print, the recounting of events is often seen as a characteristic form of oral discourse, just as the storyteller himself is seen as the prototypical creator of narrative tales. Yet here, in my own experience, the problem of evidence raises its head. Both in written and in oral cultures, narrative accounts of events, that is, not simply sequential accounts (emphasizing order) but relational accounts in a stronger sense, are less common in verbal interaction than is often supposed, less common and less elaborate.[4] Of course, visitors everywhere arrive and cannot be stopped from telling what happened on the journey; others give accounts of their experience at different moments of their lives.

An account requires the suspension not of disbelief but of discourse, at least in the sense of interplay. The narrative seeks to impose itself on conversation in a hegemonic way, but such is the nature of interaction that we rarely get the chance to finish our account. Unless we impose ourselves upon our fellows and run the risk of becoming "bores," the story often remains incompletely told, partly because its telling demands not so much an attention others are unwilling to give as an inaction they are unwilling to undergo. We can accept such a suppression of dialogue/duologue only in a formal setting—in the ritual or dramatic performance, in the sermon or the lecture, or perhaps in face of superior authority, to whom we have anyhow little to say (a professor who invites the student to tea, a queen, premier, or president who asks a subject to an audience, that is, to provide an audience). In other circumstances we as hearers allow tellers to get away with, maybe, the pithy narrative of the joke, which depends on timing and momentarily requires us to refrain from conversation. That is obviously even more the case with print. We remain silent before the written word. Although we can physically discard the novel if the narrator is a bore, nevertheless he has his day; we remain mute while "real" (or primary) time stands still so that narrative time can take over. Or better, since everything we experience is real, so that activity within the tale can temporarily exclude everything without.

I do not think this is very different among the LoDagaa of Ghana in

West Africa, whom I take as my touchstone of an oral culture, a society without writing, whose communicative patterns differ from the oral tradition within a literate one. Of course, there as everywhere, one has to suffer bores if they hold dominant positions, but in general it is dialogue rather than monologue that makes for valued intercourse, and long narrative accounts of events belong to the category of monologue. Naturally, there are situations where such accounts are called for. In dispute cases, one tries to begin at the beginning and continue to the present, or vice versa. Recounting the sequence of events is inevitably involved in the act of putting a case, since one cannot set it out beforehand in writing and take it as "read," and since the chronological order is the most "logical," the one that communicates the situation most efficiently to the mediator, arbitrator, judge, or even the anthropologist as casual spectator.

Time and "Literature" Among the "Tribal" LoDagaa

It is important to ask how prevalent the "literary" mode is in oral cultures, or rather the realm of what I shall refer to by an ugly periphrasis, "standardized oral forms," since I have doubts about calling "literary" that which is without letters. Here our attachment to a literary background may mislead. We are accustomed to thinking of standardized oral forms in terms of folktale, legend, epic, and myth, that is, in narrative terms, together with song, in which music, voice, and possibly dance are inextricably involved (in my experience the lyric does not exist as words alone).[5] All these forms are widely reported, yet I want to argue that they are not always widely used.

Folktales

Let me take first of all the most obvious case, that of folktales. These display a very definite narrative structure, although their characters are a mixture of people, animals, and gods (often God, in fact). While they are rarely specific as to absolute time, being more likely to start without any specific mention of space or time than with the sorts of temporal punctilios that open *Tristram Shandy* or *Robinson Crusoe,* they do offer the listener a sequence of tightly interconnected events for him to follow. They are short and limited in number and they are largely stories for children. It is customary for those brought up on nineteenth-century views of popular literature, fostered by German and European nationalisms, to regard such tales as the *disjecta membra* of earlier oral cultures, survivals of myths of greater significance. In fact, that is not at all the case—or rarely is. We find short tales of this kind distributed throughout the oral cultures of Africa, where they have not only a remarkably similar form and similar content but also a similar role. While my observations do not agree with the statements of some other writers, detailed examination of their accounts reveals that the context of telling has often been recalled from their youth or taken from the recollections of other people, and the stories have been recorded in specially constructed situations. But, with one exception, I have never seen adults gather round of an evening to tell such tales to one another. When I have heard them told in natural settings it has always been adults to children (into which category the visiting anthropologist easily fits), or more likely children to one another. Adults occasionally refer to such tales, but by and large in much the same way as we would refer to Red Riding Hood or Jack-in-the-Beanstalk. As for those observers who take the content of such tales as a sample of the thought of oral cultures, it is no more wonder that they end up with notions of its "primitive" nature than if some African scholar were to construe Snow White in a similar manner. Any literary form is only part—at times an important part, at times a less significant one—of the complexity of human thought and action that we call culture. It is not a part that can be taken for the whole; it is contributory, not representative.

Legends and Epics

Let me turn to the more extended forms—legend, epic, and myth—and ask how characteristic they are of oral cultures. The word "legend" itself is derived from the medieval Latin of the twelfth century and refers to that which is "meet to be read," that is, to stories of saints' lives intended to be read at matins (La Légende dorée). Later, in the sixteenth century, the word designated "a popular traditional tale, more or less fabulous," centering around real persons but embroidered by the imagination, as in the legend of Roland or of Faust. The equivalent in the oral cultures of Africa would be the clan or state histories, which we find in "tribal" and "state" societies respectively but in more fragmentary forms than is often supposed.

The legend of Roland is known not only as the Song (Chanson) but as an epic, a narrative usually in verse. The term "epic" is derived from the Greek *epos* and used specifically to refer to long, classical works such as the *Iliad, Odyssey,* or *Aenead,* whereas the more mysteriously resonant term "myth" is often applied, by Kirk for example, to the Mesopotamian Gilgamesh.[6]

All these early forms were strongly narrative, their organizational frameworks firmly grounded in space and time, albeit not to the extent of most films or novels nor perhaps to the degree demanded by the Aristotelian unities.

Owing to the work of Milman Parry and Albert Lord, the Homeric epics at least (it is not of course the same for Virgil or Milton) have been held to be oral compositions, of the same general kind as the Yugoslav song these Harvard classicists recorded in their pioneering fashion. Thus the *Iliad* and the *Odyssey* take their place next to such classic examples of the supposedly oral mode as Gilgamesh and the Rig Veda, though the latter is hardly narrative. In Henry Chadwick's theory the epics were characteristic products of the Heroic Age, the age of warrior courts that encouraged the creation of long narratives telling of fabulous struggles.

The problem with the views of Parry and Lord, and of Chadwick, is not so much locating these narratives in their actual sociohistorical context as seeing them as residual productions of oral cultures.[7] First, the discussion is often based on a failure to distinguish between the products of oral discourse, of a so-called oral tradition, in societies with writing (such as Yugoslavia) and those of oral cultures (those without writing at all). Regarding societies with writing there is consequently a tendency to treat the two traditions, the oral and the written, as distinct rather than as interacting. Secondly, the societies where the classical epics flourished were all ones with writing. Thirdly, when we look at the area that has been the largest source of compositions by oral cultures, namely Africa, we find remarkably little legend or epic, a fact noted by Ruth Finnegan in her comprehensive review of "oral literature" in that continent.[8] We find some epics in the warrior groups of the Saharan fringe, recorded by Segou, Amhate Ba, Meillassoux, and others, but these works are obviously influenced by the traditions of written Muslim culture coming from the Mediterranean. They are largely the products of professional singers (*griots*, praise singers), who adapt their versions to the particular audience whose forebears they honor.

The Myth of the Bagre

How does myth fit into this discussion from the standpoint of narrative and of embeddedness in oral cultures? I do not use the term "myth" to refer, as many do, to the cosmology or "mythology" of simple societies, reconstructed from fragmentary references in prayers, folktales, recitations, and commentaries. Instead I use it, in a simplistically concrete way, to refer to a genre, a standard oral form consisting of long, oral recitations, not fabulous in the legendary sense (though already in Greek one meaning of the word *mythos* becomes the mistaken or outworn beliefs of others), but concerned with supernatural as well as with human beings. "Myth" as cosmology does not exist as a specific discourse, the equivalent of the literary text, except of course in the writings of the observer; myth as recitation, on the other hand, is a primary construct of the actors themselves. Contrary to expectation, lengthy recitations of this kind are not common features of oral cultures. We do occasionally find them, and it is such a recitation, one I have recorded many times, that I want briefly to discuss. But their distribution is very intermittent, found in one community but not in the next, and insofar as they are characteristic of any general type of society in Africa, it is the nonheroic rather than the heroic type of society, the "tribe," not the state.

The particular "myth" or recitation in question is the Bagre of the LoDagaa of northern Ghana in West Africa.[9] Though I shift from the general to the specific, my example is not atypical of the genre. This long metrical recitation is accompanied by rhythmic beating and delivered in a chanting voice during the course of a series of ceremonies, which constitutes an initiation into an association that confers medical, ritual, and social benefits. It takes place under authoritative conditions; although the audience repeats the words of the speaker, their active participation is prevented by the formality of the setting.

There are two parts, the White and the Black, each of which takes several hours to recite. I speak here of the versions I recorded in the settlement of Birifu; neighboring communities have the same association but the recitations show many variations, more so than the actions. In the context of the uses and types of time, what strikes one about the myth is the relative lack of importance placed upon narrative, and hence upon narrative time or narrative space. There is a plot in the shape of a plan but not in the specific sense of structured narrative.[10]

Part of the reason for this lack of emphasis emerges when we think back to the "myth-and-ritual" school that flourished in classical and Near Eastern studies earlier this century. This school derived at least part of its inspiration from the work of Jane Harrison, especially *Ancient Art and Ritual,* which in turn owed much to the Frazerian influence on the classical studies of Cornford, Cooke, and others (her later work also owed something to the contributions of the Durkheimian school). Just as Robertson Smith had maintained that there is no belief without its ritual, so too Hooke and others made a similar claim about myth and ritual: the one is the verbal counterpart of the other.[11] As a general theory of myth the claim is inadequate, but it works in particular cases. The White Bagre is one of these.

Both parts are recited during the ceremonies, but the White Bagre presents an account of the rituals that the neophytes are actually undergoing. These ceremonies proceed over a period of some months, and on each occasion that one takes place, the myth is recited up to the point that has been reached in the performances. Although the White Bagre provides an account of the rituals, it is certainly not exhaustive, not a handbook for performance. Rather it is a *post hoc* recapitulation which omits much but recalls certain salient features, bringing out their importance for both the neophytes and the initiated alike. Sequential time is involved in the account: one ceremony follows the other in a set order, which is linked to the seasons partly because the provisions needed for food and beer are seasonally available, but also because the value of crops (and of the performance itself) is deliberately stressed by the imposition and gradual lifting of a series of eating and other taboos. But duration too is stressed by the many references to the time spent in the various activities, and by the repeated injunction to go round calling all the participants to come and attend in two days' time.

In the accounts of a few of the ceremonies in the series, short narrative incidents resembling folktales are embedded, their incisive plots serving again to fix those ceremonies in the minds of both speaker and listener. At the same time the recitation provides a loose framework of explanation, though here too there is little that people did not already know. The initial explanation takes the form of setting the stage at an undetermined time and place in the past when troubles were heaped upon one man, presumed to be the ancestor of all, and his family. As a result he went to a diviner and was told he had to perform the Bagre so that these problems would cease to worry him. What follows is an account both of what he then did and of what the participants are now doing; the account and the performance themselves amount to the conquest of time and the conquest of space. For the whole series of ceremonies resembles the Mass in being a recapitulation of things past, only it is less of a deliberate reenactment: what happened then is actually happening now; the ritual is not simply a theatrical mock-up. There is no progression between past and present. This is no Miracle Play representing the death of Husein, no Seder commemorating the exodus from Egypt; it is the thing itself, with the same significance as the original act. The same events occur. In one ceremony God comes down to earth; he did so then and does so now. In another the ancestors return to their old hearth and drink the beer that is being brewed for the participants.

It is the fact that these ceremonies embody replication that makes the ancestors so important. For in the absence of a Book of Hours or a Book of Common Prayer, it is old men's memories of what their own predecessors did on which the structure of authority rests. The ancestors are the learned ones who have passed on, leaving their successors to cope inadequately but as best they can (that is how the LoDagaa put it to themselves). Like God (and gods), the ancestors are both there and here, both were and are, conquering not only time but space as well.

The conquest of time does not necessarily mean the abandonment of narrative. In the prophetic religions, which are religions of the Book, the major yearly ceremonies are recapitulations of the life of the master, a life cycle rehearsed as an annual cycle, even as a weekly cycle in the case of the Mass. Nevertheless the narrative structure of the White Bagre, as recited in Birifu, is extremely weak. That is not always the case elsewhere. A version recorded by my collaborator, S. W. D. K. Gandah, in the neighboring settlement of Lawra in 1977, which we have now transcribed and translated, includes a long "prose" account of the "legend" of how the clan migrated and came to settle where it now is. In other words it describes the path of the ancestors (*saakum sor*), which is both the actual path they took (and to which sacrifices are always made in the course of the Bagre) and the path (the ways) that they followed then and that we should still follow now. Much of the imagery of the Bagre has to do with a quest, with mankind's being led along, then off, the right path, usually by the beings of the wild who are both revealers of culture and tricksters of mankind. Time in the sense of sequence (narrative) is important but time in the sense of unilinearity is not. For the journey of the ancestors results in their meeting new neighbors who teach them to perform the Bagre, a ritual they knew in the past but had since abandoned and forgotten. So in this case too the recitation about the past merges into, conforms with, present action.

This is the only example known to me of an extended clan legend of this kind among the LoDagaa. And it seems to have been elaborated specifically in the context of the Bagre, in the context, that is, of the recitation. In every village, lineage elders can give some account of the migration of the clan, and often enough one finds the story confirmed by inhabitants at the other villages mentioned. But these accounts are at best fragmentary, at worst shreds and patches, hardly narratives at all in any meaningful sense.

Time and "Literature" in the State of Gonja

The Drum

To the south of the noncentralized, tribal LoDagaa, lies the ancient kingdom of Gonja, which has been influenced by Islam and by written Arabic in a marginal way.[12] In this kingdom narrative "histories," legends, of events abound, especially around the person of the conquering hero, Jakpa, the Lord of the Towns. These accounts are highly segmentary, very different versions being given in different divisions, by different memorialists and by different ethnic groups within the state. The nearest approach to a national legend of the ruling group is the drum recital of the Kuntungkure drummers of Damongo, the present capital.[13] But this is not narrative in any significant sense. It consists of the drum titles of a series of chiefs or chiefships, the two being again confounded since praise of one's ancestors is praise of the present incumbent and vice versa. These titles are recited in archaic or "deep" Gonja, and most people cannot even understand the words, let alone the wider reference to events. Indeed one might say that the present authority of the recital in solving disputes, when the drummer is called to play and interpret, depends upon just this limited intelligibility, for the recital can be turned to support almost any cause, like the ambiguous pronouncements of other oracles, more distant in time, or the forecasting of events that are seized upon in newspaper astrology.

"The Chronicle"

In addition, there is an eighteenth-century chronicle in Arabic covering a period of some 50 years, containing comments on the significant events of each year (some dated in calendrical fashion), and beginning with a version of an earlier legend of the conquest of Gonja by Jakpa. Various copies are in the possession of the priestly Muslims of certain divisions where they are kept as a legitimizing treasure.[14] The written chronicle provides a stark contrast with the oral forms, clearly bringing out the potential consequences for "history" of even a very restricted use of writing. Nevertheless in Gonja both are treated in much the same fashion. Since very few can understand Arabic and no one the events recounted, the chronicle presents much the same situation for the community as the drum titles. But for the future historian the potential consequences are very different.

The Legends

The legends told in the various localities are different in character from the written chronicle (as well as from the drum "history"), being less structured in form and largely narrative in content, telling mainly of what happened when the Lord of the Towns first came to those parts and how he was received locally. In these oral histories little or nothing is found of later events. Virtually all is concentrated on the beginning of the Gonja state, how the rulers and their followers arrived, where they came from, whom they met. Once again the past is not altogether divorced from the present, and though we certainly cannot explain everything in these legends by reference to contemporary states of affairs, many of them incorporate a charter for the present of the kind Malinowski posited. Some of the legends are fairly gross examples of the deliberate manipulation of narrative by Gonja who found it necessary to make claims about rank and territory to the colonial conquerors after the end of the nineteenth century.[15] And today too one rarely hears of accounts of political events that do not support the status quo, unless some subordinate group, slighted dynasty, or unsuccessful prince ventures a protest.

Given the pressures of local interests, it is not surprising that the versions vary. Indeed they vary much more widely than this factor alone would suggest, telling incompatible stories about the conquering hero and those associated with him. It would be impossible for any Gonja to attempt to reconcile them, or even a sample, within the bounds of oral culture, partly because of the ambiguities about (chronological) time and space. Even when writing makes it possible to set one version beside another and to try to interdigitate events, resolve contradictions, distinguish the personae, there is little possibility of linking them into a consecutive, time-oriented document. The late paramount chief, Yagbumwura Timu II, previously J. A. Braimah, set about this task in a series of largely unpublished manuscripts, putting forward his interpretation of a number of these sources in a coherent but impressionistic fashion. Even a more rigorous literary scholarship would only serve to document variability and bring to light contradictions; it would not produce a synthetic account. It is true that early literates within Gonja society might have taken the various Jakpa stories and strung them together into some rather picaresque legend. Indeed, this seems to have been the way that a number of longer versions of epics or legends were constructed during the Heroic Age, that is, during the infancy of written literatures. But here no synthesis occurred; like the state itself, historical narrative was highly decentralized; the "overkingdom" had no overall history apart from the Gonja chronicle and the drum recital.

The Gonja, then, are greater employers of narrative forms than the LoDagaa. While they have a remarkably similar repertoire of folktales, not only is their legend more developed but so too is their use of narrative accounts in disputes that come to court. The LoDagaa were more likely to have recourse to the bow and arrow, and even now have not developed a court oratory of the kind found among their neighbors. On the other hand the Gonja have nothing of the Bagre kind, neither the White nor the Black.

Earlier I discussed the White Bagre, which we might see as the equivalent of a (written) ritual text, except that it does not act as a guidebook that serves to prompt, indeed to dominate, such performances, as is the case with religions of the Book. The Black Bagre does show a greater narrative content, even if chronological time (and even sequential time in a story-telling sense) plays only a minimal part. For the second part of the recitation expands on the notion of a quest, a seeking after truth, which in a much more simplified form constitutes the framework of the White. One of the two (first) men, described as companions rather than as brothers, sets out on a journey, during the course of which he learns, partly from God but mainly from the beings of the wild, about culture, how to perform those processes essential to human life as the LoDagaa recently experienced it. From God he learns how to procreate, from the beings of the wild how to smelt iron, grow food, make weapons, brew beer. The narrative accounts of how these tasks are carried out reflect contemporary techniques, except that the local production of iron, the description of which remains a major part of the recitation, ceased soon after the beginning of the century.

If the narrative element in this part of the myth is stronger than in the first, it revolves around these technical processes, although by no means exclusively so. The younger brother's quest after knowledge also brings him face to face with spiritual forces, with God in the Above, with the beings of the wild here on earth, and at one point with the Earth shrine, or rather with the skin (surface) of the earth (*tenggaan*). These encounters involve a discussion of philosophical issues, for example, the problem of good and evil and the question why we should observe certain prohibitions (taboos) and perform certain actions (sacrifices). Hence, while a narrative framework exists both for the Black Bagre itself and for the constituent incidents (for instance, that of the making of iron), these are not tightly structured temporally. For here too the past and the present coincide: what happened then is happening now. In the first place, the "then" is the vaguely defined time of *Genesis* rather than the more precise time of the *Chronicles*. In the second place, much of the meat of the Black Bagre, in its more elaborate versions, has to do with speculations that are foreign to the narrative mode and that bypass considerations of space and time since their significance is seen as universal. It is not so much that the LoDagaa are concerned with an original Dream Time, as has been said of the Australian Aborigines. "Once upon a time" would (if it were possible to translate) sound as childish to them as it does to us. It is rather that in ritual and myth, then and now are one and the same, or at least merged into a timeless unity.

Given this general situation, we need to ask why the myths of oral societies presented to us by many anthropologists place such a heavy emphasis on narrative, on story, on sequence. I think here specifically of the "myths," of the snippets of narrative, used, for example, in Lévi-Strauss's magisterial work on South America. There are no doubt some broad regional differences in oral style, form, and content; animals do not normally play a central part in African "myth" (as distinct from "folktale") since in the more serious, adult forms of discourse the actors are humans and their gods. But I suggest that the picture we have of the widespread predominance of narrative is partly a result of the demands and expectations of the European observer, partly of the context of "myth," and partly of the reduction of oral forms to written ones.

The demands of a European observer, coming into the society from beyond, are often precisely for a story, a narrative. Even if it were possible to recite the Bagre outside the ceremonial situation, it would be impossible to present any adequate story for the first part, and not easy for the second. Any kind of summary would be difficult, although one could easily reproduce one of the folktales that, like nodules of flint in chalk, are embedded in the recitation—for example, the tale of the guinea cock and his mate fighting over food and sex, which marks the beginning of a certain ceremony. The account of the ritual and the "philosophical" passages would inevitably be suppressed in favor of the more salient narrative, the omission giving the impression that the story, a fabulous story, is what such recitations are about. Until recently that was the dominant impression about these oral forms. The misunderstanding arose because most myths could be recited only in the ritual context. Even if that was accessible to the observer and he was very fluent in the language, it was impossible for him to write a long recitation down verbatim. Outside the context of performance, the recitation was likely to undergo considerable change and to concentrate on a summary of the narrative content, which did not approach even a précis. Only recently has the position altered as the result of the introduction of the portable cassette recorder, although in countries with more advanced technologies it has been possible, as Parry and Lord showed, to make sound recordings over a longer period.

The Telling of Time: The Oral and the Written

One of the submerged themes of this discussion has to do with the telling of time, as opposed to the time of telling. One element in the difference between narrative and the use of time relates not simply to differences between complex and simple cultures, but more specifically to differences in the means of communication. Although time was measured before writing was introduced, the shift from "natural" temporal markers (such as those in the sky) to "abstract" markers (such as those on a calendar) represents a development of great significance to science as well as to daily life, and influences not only the content but the structure of literary achievement. Neither of these changes is directly due to writing per se, but they are dependent upon the drawing of lines (a basic literate accomplishment), whether graphically or conceptually, as for example in the concept of an era, whose point of departure has to be recorded and the subsequent dates fixed by a regular written procedure. The concept of an era is not the kind of thing one either develops orally or holds in the head alone; it emerges in the interaction between man and marker.

Despite, or perhaps because of, the difficulties of measuring chronological time in oral cultures, there is often considerable internal pressure toward more exact forms of the telling of time. When I lived among the LoDagaa I was frequently asked to tell the time from my watch, whereas normally they would refer to the position of the sun; the inquiries were due not only to the prestige of my timepiece and the demands of the new schedules imposed by paid labor, but also to a more general desire for increased accuracy. It can be annoying not to meet people on the right day and truly problematic to meet them at the right hour (unlike the day, an arbitrary concept). But the desire for greater precision was perhaps more evident in the requests for me to say how many "moons" would elapse before, say, the first rains. For the people realized that my written calibration of time had a predictive capacity that did not attach to their system of trying to fit moons into the annual solar cycle (which for technical reasons is impossible), nor yet to anticipations of the appearance of a certain insect, even though, once it emerged, the latter might possibly give a more accurate prediction, being responsive to environmental factors in a way that an abstracted calendar cannot be.

But it was not only these practical uses of writing-based time measurement that were valued by the nonliterate LoDagaa and their neighbors. In the account books of an interesting twentieth-century Vai figure in Liberia called Ansumana Sonie, who made extensive use of the locally invented script, Michael Cole, Sylvia Scribner, and I found he had written down in Arabic numerals the exact dates of events, especially birth dates, in the manner that was common in European Bibles.[16] There was no particular reason for this, especially since the widespread use of the time of birth for astrological purposes in the Chaldean-derived traditions that spanned the Old World from the Blasket Islands to the Philippines was not, to my knowledge, found in this peripheral corner of Islam. Sonie used birth and subsequent dates to play with in unexpected ways, calculating elapsed time between events. Part of his interest in so doing may have derived from the preoccupation of the surrounding written cultures with dates of birth, which were used to calculate age and to mark birthdays. Birthdays soon became a preoccupation with schoolchildren in the Third World, even where no birth registration existed and no traditional ceremony was observed. Date of birth, along with forename and family name, became an essential feature of a card of identity, and was then inscribed on the tombstone. The birthday tea party for children was one of the major festivals adopted by the African bourgeoisie in the 1960s in Ghana to bring together school friends and neighbors, as distinct from kin. It was a strong but perhaps unintentional marker that distinguished them and us from their brethren in the oral culture.

What I am suggesting here is not simply that the advent of writing makes a difference to time measurement, to the recording of events, and to people's conceptions of time (a well-established theme) but that oral cultures are often only too prepared to accept these innovations. The native Americans, even with their simple system of pictographic protowriting, achieved the recording of annual events in, for example, Lone Dog's Winter Count among the Dakota. Such forms of time recording may already have been stimulated by the European invaders. But in any case, the important role of the calendar in Central American graphic systems provides further irreproachable evidence of the early tendency of writing systems to develop chronological chronotypes, more exact and exacting modes of reckoning time.

Time and Narrative in Oral and Written History

All this bears upon the emergence of history as we know it and its differentiation (at least the differentiation of the narrative history recently espoused by Lawrence Stone) from other narrative forms and more generally from wider concerns with the passage of time. Concepts of time as well as the implications of time are always there; people look back at their personal as well as their "collective" past, in the latter case both genealogically and in relation to events (*l'histoire événementielle*). But the selection of kin and events is certainly more dominated by the present in oral societies than in those where genealogy or events have been recorded in writing, so that in oral societies the notion of the past as a charter for the present carries more (though not exclusive) relevance. Genealogies, even when written, cannot themselves be regarded as history since, although they embody sequence, they do not entail consequence. Indeed somewhat the same can be said of chronicles of the kind found in Gonja, which incorporate measured time but again not consequence, not narrative, but simply recording. Clan and state legends on the other hand do utilize narrative, often in a relatively tight sense, at least with regard to movement through space; time is inevitably more difficult to be precise about.

As others have argued, the oral genealogy is often more a statement about space (especially social space) than it is about time since among other things it often represents the contemporary, or near-contemporary, distribution of living groups as much as the relations between dead individuals. Such a use of the past is of course also known in written history, but the difference lies, importantly, in the matter of degree and in the fact that the written raises in an acute form problems of the intercalation of the past and the present, the written and the oral, and the nature of views of the past and their value for the present. The opposition between the meanings of oral and written genealogies was dramatically brought out in the use of the Tiv genealogies in colonial West Africa of the twentieth century and in the use of the Domesday Book in colonial England under the Norman invaders.[17] In the first example conflict arose between the colonial administrators and the Tiv "tribesmen" over the oral genealogies written down by the colonialists 50 years before, which no longer accorded with the local views of the "past" that were so highly conditioned by present concerns. So too with the Domesday Book; a bookkeeping of conquered England written at one instant in time gave rise to conflict when it continued to serve as a point of reference for law courts for the next few centuries.[18]

I do not mean that there can be no accounts of the past of oral cultures either by observers or by the actors themselves. Both however differ from accounts based upon written records, which, following the usage adopted by archaeologists and other scholars when they speak of prehistory, I refer to as history or graphohistory. One can of course use *history* as an inclusive term for all kinds of accounts of the past and put the same term in quotes to refer to literate chronicles. But that solution seems cumbersome.

The relative absence of histories of and within oral cultures has nothing to do with any "childlike" nature. It is rather a question of a lack of means. It may also be a question of the lack of interest (the two are sometimes closely intertwined), but I am not happy with that notion since one of the first scholarly activities among members of newly literate groups in contemporary Africa is to write their own histories; this is of a piece with the pressure from below for greater precision in the measurement of time. This activity is undoubtedly promoted by contact with the historiographic traditions of Islam and Europe. At the same time, the impulse seems more widespread than such specific contacts imply. And it is not a question of people having history in their heads waiting to be written down. New cognitive tools enable them to undertake new cognitive tasks. To speak about their historicity or sense of history in an unrefined way is to abandon important distinctions; Derrida notwithstanding, to be able to "read" the stars in an oral culture does not have the same implications as to be able to read a (written) text. To extend concepts metaphorically in such an overly relativistic way is to meet the "other" not face to face but back to back; to force our history, our reading, on them is to abandon history, to see development and evolution (in all their complexity) as stopping with the apes. We are different, but are we so different? What we need is neither oppositions nor identities but discriminations, discriminations which are contextualized by the questions we ask and by the tasks we undertake.

There is a similar but not identical problem with narrative. I am not suggesting that the narrative form is totally alien to oral cultures, only that it is less common than is often supposed, and that what does exist is more loosely structured than in cultures with writing. It is difficult to measure the tightness or looseness of narrative, but I suggest that the following simple criteria are relevant: (1) the preciseness of the location in time and space (i.e., the presentation of background); (2) the development of characterization; (3) the proportion of non-narrative material; (4) the integration of the plot, that is, whether segments are substitutable or instead interlocked, and to what degree.

The claim that the "literature," the standard verbal forms, of oral societies displays a relative lack of emphasis on narrative partly arises from the fact that these genres are differently constituted. "Myth" in the shape of the Bagre (and elsewhere) is not mainly a narrative; it does a great deal more work that for Western audiences would be allocated to other written genres, not simply literary ones in the restricted sense. The question is not simply one of exclusive genres but of an exclusive conception of the "literary" and the other. That conception represents not only a two-world division between "fiction" and "reality" but also an implicit notion that "literature" is the equivalent of the "literary." This notion leads to a neglect of other written materials, the equivalent of which in the undifferentiated or less differentiated oral world would form part of more inclusive genres. The Bagre is philosophy and religion as well as narrative.

I took as my starting point the treatment of time typified in Defoe's *Robinson Crusoe,* which was consequent on changes in the mode of communication, namely, printing, as much as on changes in the mode of production, namely, capitalism. This view of time I contrasted with that in two societies in Africa, the tribal LoDagaa, where there was no writing until recently, and the state of Gonja, which was peripherally affected by Islam. There were a few broad differences in the standardized oral forms, the oral genres, of these two societies: the state was more concerned with legend, with the Arabic notion of an era, and with an Arabic chronicle dating from the eighteenth century, while the tribe possessed a long recitation, the myth of the Bagre, a secret society found among the LoDagaa. However, even that extensive recitation had only a minimal chronological component. Despite the presence of short folktales with a structured frame, narrative seems to be less common in "oral literature" directed to adults than is often thought. Like history and chronicles, narrative came into its own with the acquisition of writing, which facilitated and promoted a more specific treatment of dating, of duration, and even of sequence, both for the telling of time and for the time of telling.[19] I stress specificity because the widespread and rapid adaptation to new modes of reckoning suggests an already-existing pressure to build on the linear patterns rather than remain wholly satisfied with the circular formations in which so much of oral culture is embedded.

PART III

Time and the Politics of Criticism

Time and Timing: Law and History

GAYATRI CHAKRAVORTY SPIVAK

"TIME" IS A WORD TO which we give flesh in various ways. Kant taught the European that he could not be or think or act without this first gesture.[1] Freud unhooked this lesson from its easy reading—the primacy of real lived time as giving us life itself—by suggesting that "real lived time" is produced by the machinery of the mental theater.[2] Even this is a bold new technique of fleshing out that word "time." In this essay, I will assume that one common way of grasping life and ground-level history as events happening to and around many lives is by fleshing out "time" as sequential process. This I have called "timing." I will assume, further, that this feeling for life and history is often disqualified, for the sake of a dominant interest, in the name of the real laws of motion of "time." This version of "time" I have called "Time." It is my contention that Time often emerges as an implicit Graph only miscaught by those immersed in the process of timing. I have sketched this tyranny of the "visible," of the "good writing," in a text of Hegel on the *Srimadbhagavadgita* as well as in the *Srimadbhagavadgita* itself.[3]

Two further points need to be made here.

First, the tyranny of the visible over the merely lived covers over the intermingling of spacing (*espacement*) and timing in the "lived." This major argument of deconstruction will not enter this essay, but only the book of which this is a part.[4]

Secondly, to notice such a structural complicity of dominant texts from two different cultural inscriptions can be a gesture against some of the too-easy West-and-the-rest polarizations sometimes rampant in Colonial Discourse Studies. To my mind, such a polarization is too much a legitimation-by-reversal of the colonial attitude itself. This second argument is the backbone of my essay.

The usual political critique of the Hegelian dialectic is to say that it finally excuses everything.[5] Another way of putting it is that because Hegel places all of history and reality upon a diagram, everything fits in. Thus, in the Hegelian pictures of the journey of the *Geist* given in *The Philosophy of History, The Philosophy of Right,* and *The Lectures on the Aesthetic,* the laws of motion of history are made visible as, concurrently, the Hegelian morphology is fleshed out.[6] The time of the law has the spaces of a rebus, the active reading of which will produce the timing of history. In Hegel's own words, "the intelligible [*das Verständige*] remains [*stehenbleiben*] in concepts in their fixed determinateness and difference from others [*von anderen*]; the dialectical [*das Dialektische*] exhibits them in their transition and dissolution."[7] As a literary critic by training, I will concentrate on a couple of paragraphs from the *Aesthetics*. Because I am Indian and was born a Hindu, I will also attempt to satisfy the increasing and on occasion somewhat dubious demand that ethnics speak for themselves by focusing on a bit in Hegel on Indian poetry.

According to Hegel, there are three moments in a work of art: the form, or *Gestalt*; the content, *Gehalt* or *Inhalt*; and the meaning, or *Bedeutung*. The true meaning, not only of a work of art, but of any phenomenal appearance, is the situation of the spirit on the graph of its course toward "self-knowledge." (This too is basically a graphic intuition: more the spirit and knowing-it exactly coinciding when superimposed, as it were, than the spirit, in a subjective model, "knowing itself.") Starting from a situation where content and form are intertwined in an unacknowledged unity as meaning, the elements must separate with some violence so that "conscious" conciliation between spirit and knowledge may finally be effected. At that stage of adequate superimposition or "identity," since there is no separation between sign (content/form) and the transcendental meaning (spirit-in-self-knowledge), there is no art. "Art" is the name of the sign of the lack of fit between the two graphs—spirit and its knowing. It is well known that the spirit or *Geist* that acts out the scenario of self-knowledge is not something like a grand individual subject. It is rather like the principle of subjectivity, in other contexts given a world-historical nuance.

What we have in Hegel's narrative of the development of art forms, then, is not an epistemology, an account of how an individual subject (or subjects) knows or knew and produced commensurate art, but an epistemography, a graduated diagram of how knowledge (an adequate fit between sign and varieties of meaning) comes into being. Art marks the inadequacies on the way. It is a dynamic epistemograph since the emergence of the finally adequate relationship between sign (spirit) and meaning (knowledge) is the result of much straining on the part of both to achieve a fit. Each new configuration steps forth in the sublation of the earlier stages of the struggle. The "deviations"—lack of fit—on the way are therefore teleologically "normative."[8]

Upon this epistemograph—a graduated diagram of the coming-into-being of knowledge—the art of Persia, India, and Egypt are not granted the status of being produced *by* the spirit, however unfitting such art might be to the graph of true knowledge. They are all normative deviations in the area of the *un*conscious symbolic. The task of the Hegelian philosopher of art is to decipher this epistemograph. The relationship between form and content in this art can only be evidence of a struggle toward signification; it cannot be an intended collective sign of a stage in the journey toward adequation.

By the time we get to India, the shape (*Gestalt*) is "perceived" (by the *Geist* as subject, not Indian individuals) to be separate from the meaning. Indian art seeks to give an *externally* adequate representation, according to Hegel, to the grandeur of a meaning that is perceived as beyond phenomenality. Thus, unlike the scenario as run by the proper inner process, Indian art cannot supersede or sublate the contradiction between shape and meaning. The contradiction "is supposed to produce a genuine unification . . . yet," in Indian art, "from one side it is driven into the opposite one, and out of this is pushed back again into the first; without rest it is just thrown hither and thither, and in the oscillation and fermentation of this striving for a solution thinks it has already found appeasement" (*LA* 1: 333–34). Therefore, "the Indian knows no reconciliation and identity with Brahma [the so-called Hindu conception of the Absolute] in the sense of the human spirit's reaching *knowledge* of this unity" (*LA* 1: 335). (Who this "Indian" might be is of course an irrelevant question here.)[9]

The *verstandlose Gestaltungsgabe,* translated by T. M. Knox as "unintelligent talent for configuration," that Hegel sees as the *Geist*'s normative aesthetic/epistemic representation in this *static* "Indian" moment on the chronograph spans a few millennia—from at least the second millennium B.C. to the fifth century A.D.—embracing scattered examples from the *Vedas,* the fantastic cosmogonies of the *Puranas,* the *Srimadbhagavadgita,* and Kalidasa's play *Sakuntala* (the last translated by Goethe). By predictable contrast, Hegel provides a detailed account of the various stages of Christianity and a careful distinction between Greece and Rome.

Of Hegel's "Indian" readings, I have chosen his comments on two passages on the *Gita* because they dramatize most successfully my thesis that Time graphed as Law manipulates history seen as timing in the interest of cultural-political explanations.

Hegel quotes two rather beautiful passages from the *Gita*. By contrast with the deeply offensive passages about Africa and history in *The Philosophy of History,* for example, the tone of Hegel's comments is ostensibly benevolent.[10]

> So it is said, e.g., of Krishna . . . [*sic;* Hegel uses the genitive, but it should read "by"]: "Earth, water and wind, air and fire, spirit, understanding, and self-hood are the eight syllables of my essential power; yet recognise thou in me another and a higher being who vivifies the earth and carries the world: in him all beings have their origin; so know thou, I am the origin of this entire world and also its destruction; beyond me there is nothing higher, to me this All is linked as a chaplet of pearls on a thread; I am the taste in flowing water, the splendour in the sun and the moon, the mystical word in the holy scriptures, in man his manliness, the pure fragrance in the earth, the splendour in flames, in all beings the life, contemplation in the penitent, in living things the force of life, in the wise their wisdom, in the splendid their splendour; whatever natures are genuine, are shining or dark, they are from me, I am not in them, they are in me. Through the illusion of these three properties the whole world is bewitched and mistakes me the unalterable; but even the divine illusion, Maya, is my illusion, hard to transcend [*duratyaya,* difficult to cross]; but those who follow me [shelter in me] go forth beyond illusion [*mayametam taranti,* cross over this illusion]."[11] Here [the] substantial unity [of the formless and the multiplicity of terrestrial phenomena] is expressed in the most striking way, in respect both of immanence in what is present and also the stepping over [*hinwegschreiten*] everything individual. In the same way, Krishna says of himself that amongst all different existents he is always the most excellent: "Among the stars I am the shining sun, amongst the lunary signs the moon, amongst the sacred books the book of hymns, amongst the senses the inward, Meru amongst the tops of the hills, amongst animals the lion, amongst letters I am the vowel A, amongst seasons of the year the blossoming spring," etc. (*LA* 1: 367)

However benevolent or admiring Hegel's remarks might be, they still finally insist that these passages from the *Gita* illustrate the mindless gift for making shapes [*verstandlose Gestaltungsgabe*] and an absence of the push into history.

Obviously, Hegel has to quote lists because he needs to say that the Spirit-in-India makes monotonous lists in a violently shuttling way. Hegel's conclusions from these rather difficult passages can be summarized as follows: the recitation of the height of excellence, like the mere change of shapes in which what is to be brought before our eyes is always one and the same thing over again, remains, precisely on account of this similarity of content, extremely monotonous, and on the whole, empty and wearisome.

Now, the alternative to accepting Hegel's reading is not necessarily to propose a reading that would pronounce the *Gita* politically, philosophically, or yet aesthetically correct, profound, and fine. One constructive alternative is, I think, to gain enough sense of the text and its place within a historical narrative to realize that the *Gita* itself can also be read as another dynamic account of the quenching of the question of historical verification. In fact, such a sense of the place of the *Gita* within a historical narrative is provided by its setting within the epic *Mahabharata*. The *Gita* is a tightly structured dialogue in the middle of the gigantic, multiform, diversely layered account of the great battle between two ancient and related lineages. Here the battle is stalled so that the merely human Prince Arjuna can be motivated to fight by his divine charioteer, Krishna. All around the *Gita* is myth, history, story, process, "timing." In the halted action of the text is the unfurling of the Laws of Motion of the transcendence of timing, the Time of the Universe. The *Gita* too substitutes immanent philosophical significance in the interest of a political intervention where killing becomes a metonym for action as such.[12]

I should like to distinguish my approach from two that I have chosen as representative of innovative or reconstellative readings of the *Gita*. One is used in D. D. Kosambi's "Social and Economic Aspects of the Bhagavad-gita," an essay from which I have already quoted (see notes 9 and 12). The other is developed in Bimal Krishna Matilal's unpublished work on the study of contemporary Indian cultural formation through the Indian epics.[13]

For the general reader, Kosambi's essay remains the best guide to the non-exemplary character, indeed the elusiveness, of the *Gita* in its "appropriate historical and geographical context." He establishes the peculiar contradictory interpretability of the *Gita* and concludes: "The *Gita* furnished the one scriptural source which could be used without violence to accepted Brahmin methodology, [as also] to draw inspiration and justification for social actions in some way disagreeable to a branch of the ruling class. . . . It remains to show how the document achieved this unique position."[14] His answer: "the utility of the Gita [*sic*] derives from its peculiar fundamental defect, namely dexterity in seeming to reconcile the irreconcilable."[15] The way to this answer, apart from laying out the expedient ambivalence of all overt idealisms, is, for him, through realistic and characterological narrative analysis.

My goal, by contrast, is specific to my pedagogic-institutional situation. I repeatedly attempt to undo the pious opposition between colonizer and colonized implicit in much colonial discourse study. Therefore I must show that there are strategic complicities between Hegel's argument and the structural conduct of the *Gita*. I also make an attempt to fill the empty place of the discourse of the colonized, in however imperfect a way, and suggest a method appropriate to Departments of English or Cultural Studies, not the expert historian of India. By contrast to Kosambi's realism and emphasis on character study, therefore, my way is to point out the moves in the *structure* and *texture* of the text—a text that is performative in the sense of being an island of *diagesis* in a sea of *poiesis*—that will persuade the assenting reader.

Matilal's work attempts, among other things, to deconstruct the opposition between colonialists and nationalists, as well as between developmental realists and mystical culturalists, by pointing at what he perceives to be a "dissident voice" within the text. Again, this new politics of reading may be useful in the Indian context. In the disciplinary scene of British philosophy (Matilal teaches at Oxford) Matilal's new work relates to the ethical arguments within analytic philosophy as spelled out by writers like Bernard Williams or Thomas Nagel, whose positions are unremittingly Euramerican. (My disciplinary placing, as far as I can understand it myself, I have already sketched.)

The *Srimadbhagavadgita*, the full name of the text, is a considerably later dramatic narrative addition to the epic *Mahabharata*.[16] The rest of the immense poem is ostensibly sung by the poet Vyasa. This bit is sung by God, the graceful lord—that is the meaning of the title. The short title simply means "sung," but implies, of course, the full designation, where the subject is so powerful that it cannot be actively forgotten even when absent. The "intent" of this addition to the epic is clearly to anagogize the political. It is a text composed for interpretation (and therefore designated as one of the "Vedantas," wrenched out of its "appropriate context").[17]

It is not my design to occupy the subject position of the privileged interpreter. I should like to construct a crudely "dialectical" reading of the actual narration of the *Gita* in terms of the play of law and history. Had Hegel the wherewithal to read it this way? I think so. The reading that I am going to propose is considerably less complicated than, say, the celebrated reading of *Antigone* in the *Phenomenology* and requires no more knowledge of the "Indian background" than Hegel himself professed to possess. It requires merely an impossible anachronistic absence of the ideological motivation to prove a fantasmatic India as the inhabitant of what we would today call the "preconscious" of the Hegelian Symbolic.[18]

Because "Hegel" (the name is a world-historical metonym here) wants and needs to prove that "India" is the name for this stop on the spirit's graphic journey he makes his "India" prove it for him. Such moves are not unusual among the ideologues of imperialism. Yet, given the far-reaching hold of the Hegelian morphology, it is historically impossible to "reject" it effectively where Anglo-American-style tertiary education is possible. Indeed, it may well be disingenuous or irresponsible to offer the philosophizing in the *Gita* as a competing "ethnophilosophy" evincing a tacit performative acceptance of a European disciplinary framework. It is more effective, I think, to critique Hegel from within, to turn Hegel's text away from itself, to notice that the so-called time-bound bits are crucial to the system, as does Derrida in the staging of Hegel-in-the-fetish.[19]

Indeed, there can be no correct *scholarly* model for this type of reading, of which my reading of the *Gita* is an example. It is, strictly speaking, mistaken, for it attempts to transform into a reading-position the site of the "native informant" in anthropology, a site that can only *be* read, by definition, for the production of definitive descriptions. It is an (im)possible perspective.[20] This is not the plausible perspective of a Hindu contemporary of Hegel's bemused by Hegel's reading. (In fact, a few decades later, the slow epistemic seduction of the culture of imperialism will produce modifications of the *Gita* that argue for its world-historical role in a spirit at least generically though not substantively "Hegelian." And these will come from Indian "nationalists.")

Instead, this perspective makes up the figure of an implied reader "contemporary" with the *Gita*. (This gives him—gender advised—a span of some two centuries to float about in, still less than the Hegelian arrested space of India on the graph.) Such a reader or listener acts out the structure of the hortatory ancient narrative as the recipient of its exhortation. It is highly improbable that any contemporary reader or listener behaved quite this way. But if he had, he would have been bemused that a text that was the site of the most obvious negation and sublation of history (if he can think English he can be imagined to think Hegel) should be adduced as proof of eons of ahistoricity. Moreover, Hegel himself and many present-day readers of the exotic literatures of the past did and do assume such an implausible, if often unacknowledged, contemporary reader: contemporary with the text, and far removed from "our" time. My own purpose in evoking such an improbable perspective is of course not to accept the centralized interpellation to *be* a native informant. Rather, as a teacher I am calling for a critic or teacher who has done enough homework in language and history (*not* necessarily specialist training) to be able to produce such a "contemporary reader" in the interest of active interception and reconstellation. This is different from teaching the producers of neocolonialist knowledge to chant in unison, "One cannot truly know the cultures of other places, other times," and then diagnose the hegemonic readings into place.

It is interesting that both Kosambi and Matilal presuppose the figure of such a reader: "The lower classes were necessary as an audience, and the heroic lays of ancient war drew them to the recitation. This made the epic a most convenient vehicle for any doctrine which the brahmins wanted to insert."[21] And Matilal: "Perhaps the historian has to eavesdrop on the dialogue between the past and the then present of earlier times."

The implied receiver of the exhortation in the text is Arjuna, the prince unwilling to kill his relatives in battle. The sender of the exhortation is Krishna, not only god-turned-charioteer, but also prince of a house not included among the two main contenders in the battle. There is a crucial moment in the nearly unbroken exhortation where the narrative makes Arjuna ask the question of history in the simplest way. In search of evidential verification from history as sequence or timing for Krishna's transgression of the historical, Arjuna asks Krishna: "Your birth was later and the birth of the sun was earlier. How should I know that you said all this first?" (This version of the question of history, asked within the story as performed, must be strictly distinguished from the question of historicity, which seeks to establish the locatable truth-value of the story as illocution.)

The "all this" in question is the third canto of the poem, Krishna's long lesson on how to act knowingly but without desire. Arjuna's question is placed at the beginning of the fourth canto, to provide an opening for Krishna, to give him a chance to clinch the lesson of canto 3, to speak of renunciation through knowing action. The question is ostensibly provoked by Krishna's claim that he had told this unchanging way of knowing (lawful) action to the sun.[22]

It is quite appropriate to bring up the question of history here. Krishna is not offering a more primordial mode of being where time has not yet been caught in the thought of sequentiality. In fact, Krishna's claim traps heliocentric time into genealogical time through the mediation of the law. The Law in this case is Krishna's secret passed on by the immutably law-abiding sun to the mythic human law-encoder Manu. That is the substance of Krishna's speech, and the ostensible reason for Arjuna's question, as presented by the text. Manu relays the secret to Ikshvaku, the eponymous progenitor of the Sun dynasty, in which "Sun" has become an honorific proper name for the best genealogy of kings. Thus it is proper here to presuppose a certain connection between truth and history-as-timing and ask, "How shall I know (the truth of) what you say? How should I verify what you say, since you came after?"

To this Krishna gives three kinds of answer, which prepare for the subordination of history as timing to law as the graph of Time. The first two are as follows:

(a) We come and go many times. I know all the turns. You do not. Krishna is the proper agent or the perfect subject of knowledge. One cannot ascertain sequential verification by means of just *this* history.

(b) I become by inhabiting my own nature through my own phenomenal possibility, although I am not born but am of immutable [the epithet is explained in note 22] spirit and am the head of all beings (who have already been).

I hope it is obvious from my carefully awkward translation that heavy philosophical issues are entailed in (b). An informed discussion of such issues is irrelevant to the figuration of the perspective named "native informant." Let us simply notice that human historicity is shown here to be of limited usefulness as an explanatory or verificatory model. For here the privileged or exceptional subject of knowledge is also claiming to be the subject of exceptional genesis by a self-separated auto-affection. The divine male separates itself from itself to affect a part of itself and thus create. What in the human male would be nothing more than the dead inscription of spilled seed becomes, in God, self-origin and self-difference.[23] Nature (*prakrti*) in (b) is already available as the female principle (as well as roughly the two most common senses of "nature" in English) over against (specifically male or phallic) "man" (*purusha*). The word I translate as "inhabiting" (*adhisthana*) does carry the sense of "properly placed," as a *genius loci* is properly placed in its *locus*. And if the self-generating subject properly inhabits the female in itself in order to become, the instrument is his own *maya*. I have translated this word as "phenomenality" but in the *Gita*'s Sanskrit it already carries the charge of "illusion," as indeed does *Schein* in *Erscheinung*, the German word most commonly translated as "phenomenal appearance." Working with the metaphors that hold the metaphysics here, rather than merely conceptualizing the allegory, one could say that historical verification by temporal presence is being dismissed not only by Krishna's statement that the human being is present many times around, but further by adducing the proposition that when *I* am present it is by a mechanism different from any other. *I* give the *logos* outside of historical temporality because *I* carry the phallus outside of physiological obligations.[24] Our native-informant-cum-contemporary reader would not have this specific vocabulary, but Hegel would.

Krishna's third type of answer:

> (c) I make myself whenever the Law is in decline.

The three-part phallogocentric negation and sublation of history represented in (a)-(b)-(c) can be grasped easily. Yet even such a sublation of history as timing through the mediation of law—the vanishing moment of sequential human temporality into a catachresis named Time—is not the final hortatory instrument of the text. Offering a structural summary of a highly repetitive exchange, let us say that Krishna, the privileged and exceptional subject of Time, withdraws into mere human timing and the arena of history by way of a staging of the indulgence of acknowledged error. We move toward this in canto 10, where Krishna inserts himself into *one* model of sequentiality, if not the temporal, by describing himself as the best of a bewildering number of discontinuous series. (This is one of the passages Hegel quotes as simply a monotonous repetition of what goes on in millennia of Indian aesthetic representation.)

In canto 11, Arjuna's reaction to the entire "transcendental" or "exceptionalist" suprahistorical first part of the narrative is one of acknowledged error and a prayer for indulgence:

> evam etad yathattha tvam ātmānam paramesvara
> drastum icchami te rupam aisvaram paramesvara.

The strongest burden of this couplet is the most emphatically implied "yet" between the two lines. The first line says, "Yes, Lord, you *are* as you say": by the mechanics of transcendental nonrepresentability you *are* the holder, and the singular example, of a special law. The second line says, "I want to see this omnipotent, omnipresent, infinitely wise form." The relation between the two lines is, "Sorry, I know it's wrong (a category mistake? lack of faith? human frailty?) but . . . " It is in response to this important self-excusing request that the text stages Krishna showing himself as cosmograph and indeed, in a peculiar way, as an ontograph that can contain a historiograph. (This is the other passage that Hegel quotes as proof of the monotonous repetition of the same monstrous representation in millennia of static "Indian" aesthetic representation.) Apparently to indulge the history-bound human insistence on timed verification, here in a somewhat unreally prolonged present, Time as exceptionalist graph must be negated into this more vulgar graphic gesture (Krishna's showing himself)—the famous *viswarūpadarṣana* (the vision of the universal form) in the *Gita*.[25]

Let me explain the abominable neologisms in "an ontograph that can contain a historiograph." When, in response to an unendorsable request, Krishna shows himself to Arjuna as containing the universe, he must also expand the dimensions of his own body (I am aware that this is an epic *topos*): he appears with a thousand arms and a thousand eyes, and so on. Hegel dismisses this proliferation as "monstrosity without aim and measure" (*LA* 1: 338), when in fact it is a cultural idiomatic ruse in the dialectic between law and history. But of much greater interest to me is the move that makes Krishna *contain* all origins, all developments, *and* also the present moment.

Here is Arjuna in the battlefield. He is watching the two sides. *There* are his own people—*there* are his cousins on the other side. "All these sons of Dhrtarastra [Arjuna's uncle, the father of his enemies]," Arjuna is presented as saying, "With hosts of kings, Bhisma, Drona, [Karṇa] the *Suta*-son, as well as our own chief warriors, are hastening into your terrifying and tusky mouth. Some can be seen sticking in the gaps between your teeth with their heads crushed to powder." This vivid and memorable passage is a description of the actual phenomenal present in which Arjuna is standing. He is *seeing* an alternative version of *this* Krishna chewing up all *these* people in his mouth. Krishna as a graphic representation of (a) transcendental Being (ontograph) contains the fluid present-in-time (historiograph). No explanation is needed here: the graph is evidence, as required.

The human agent in his present-in-time (his here and now) can no longer trust the here and now as the concrete ground of verifiability. It is structurally most appropriate, *and* a support to the hortatory power of the text for the "contemporary receiver," that Arjuna now speaks as follows to Krishna, a person who had hitherto been his friend. These most moving lines are an apology for action justified by the phenomenality of mere affect: "If, thinking you friend [*sakheti*], I have too boldly cried, hey Krishna, Yadava [almost a patronymic], friend [*sakheti*]; and if, through ignorance of this greatness of yours, or through sheer love or absence of mind, I did wrong [*asatkrta*] for the sake of fun—on walks, in bed, sitting or eating, alone and in company—since you are boundless, forgive me, I beg you."[26] Through the grotesquely phenomenal representation (*by Arjuna*) of Krishna masticating the details of the immediately perceptible phenomenal reality in time and space, the authority of the here and now is undermined, and in the reaction (*by Arjuna*), the phenomenality of affect is denied and produced as excuse. (It is to be noticed that, in the first line of the quoted passage, Arjuna uses *sakheti,* "as (if) a friend," twice: once as an adverbial phrase modifying *yaduktam,* "whatever I said," and once as a mode of address, a noun in the vocative case, "you who are *as if* a friend." It is all the more noticeable because the second occurrence is, strictly speaking, grammatically incorrect and unnecessary, or merely semigrammatical. Here the Jakobsonian poetic function—"as if friend" repeated twice apparently for the sake of symmetry—underscores the illusoriness of judgments based on the phenomenality of affect.)[27]

Why is this graphic sublation (negation and preservation on another register) of the apparent phenomenality of lived time and affect performed in this poem? Again doing injustice to a complex and repetitive text, I would propose that it is performed in the interest of the felicitous presentation of a *concrete* social order within a frame that has now been disclosed as an indulgent allowance for human error. This section of the *Gita* is not much celebrated in the current conjuncture.[28] My suggestion is that it is in this section that the actual *social* exhortation comes, framed not as a betrayal or contradiction of the abundantly celebrated transcendental sections but as an appropriate concession, an acknowledgment of human error, an indulgence. The tone of the narrative becomes much more "temporal" (to use that charged adjective) after this.

It is through these cantos, then, that the four castes—*Brāhman*, *Ksatriya*, *Vaisya*, *Sudra*—can be named as such:

> Control of mind and senses, austerity, purity, patience, uprightness, knowledge, insight, and belief in a hereafter are born of the proper being [*svabhāva*] of the *brāhman*. Prowess, energy, perseverance, capability, steadfastness in battle, gift-giving, and feelings of lordship are born of the proper being of the *ksatriya*. Agriculture, cattle-herding, and trade are born of the proper being of the *vaisya*. The proper being of the *ṣudra* generates work whose essence is to serve others.[29]

The happiness that is proper to the being named *ṣudra* is elaborated thus: "the happiness which, first and last, arises from the confusion of sleep, sloth, and delusion."

These vignettes are far indeed from transcendental graphing. They are customarily taken to be proof of the functional heteropraxy of Hindu social behavior. My point is that, in this authoritative text, taken as static and monotonous by Hegel, such summaries are allowed through a textual ruse showing the self-excusing errant request endorsed as divine indulgence of human error, and the subsequent disavowal by the historical agent of both the phenomenality of affect and the grounding verifiability of the so-called concrete lived present. The proper name of the caste stands as a mark to cover over the transition from a tribal society of lineage, where one cannot kill one's own kin, to something more like a state where one's loyalties are to more abstract categories of self-reference.[30]

Through this crudely dialectical reading of a moment in the *Gita,* I have attempted to deconstruct Hegel's graphic self-differentiation from the subject in India (*one* stage of the unconscious symbolic). I have attempted to show that "Hegel" and "the *Gita*" can be read as two rather different versions of the manipulation of the question of history in a political interest, namely, the apparent disclosure of the Law.

One of the differences is that the *Gita* is exceptionalist whereas Hegel is normativist. I have suggested elsewhere, in the context of gendering, that exceptionalism might be one part of the Indic regulative psychobiography.[31]

In an early passage in his critique of Hegel, Marx writes in a way that is coherent with a graphic image of the Hegelian system. If the orchestration of the marxian passage in its context is attended to, I believe it can then be seen as suggesting that that system makes appear the Being estranged from itself (*sich entfremdetes Wesen*—Being not adequate to its own proper outlines, as it were) even as it seems to present Being coming home to itself through a process of necessary othering and sublating (*Entäusserung/Aufhebung*): "It is the confirmation [*Bestätigung*] of apparent being or *self-estranged* being in its negation [*Verneinung*] . . . and its transformation into the subject."[32]

An interesting reading is produced if Marx's use of *Verneinung* here is related to Freud's later use of the term (with which it is not inconsonant), now often translated as "denegation": "A negative judgement [*die Verurteilung*] is the intellectual substitute for repression; its 'no' is the hall-mark [*Merkzeichen*] of repression, a certificate of origin—like, let us say, 'Made in Germany.' "[33] By this line of reasoning, the judgment becomes a visible graphic mark of the negation.

If Marx is read retrospectively by this line of reasoning, the Hegelian graph makes visible the repressed certificate of origin: "Made in (or for—

effect or condition) Capitalism." Marx shows this by shifting the system to "the sphere of political economy" to show up the estrangement of the system, its derailment, so that the results computed by it are reversed to the point of irrelevance: "In the sphere of political economy [the] realization [*Verwirklichung*] of labour appears as a *loss of reality* [*Entwirklichung*] for the worker, objectification [*Vergegenständlichung*] as *loss of and bondage to the object* [*Verlust und Knechtschaft des Gegenstandes*], and appropriation [*Aneignung*] as *estrangement* [*Entfremdung*], as *alienation* [*Entäusserung*]."[34] According to Freud: "By means of [*vermittels*] the symbol of negation, thinking frees itself from the restrictions of repression and enriches itself with material that is indispensable for its proper functioning." It is perhaps in acknowledgment of an enriched and proper functioning that still owes something to denegation (in a proto-Freudian rather than a strictly "philosophical" sense) that, in the sentence that follows my quotation, Marx dockets the Hegelian system in the narrative timing of *das Aufheben*—the effort of sublating—rather than the graphic Time of *Aufhebung*, the accomplished sublation.

One of the most scandalous examples of such a slippage between the effort of sublating and accomplished sublation is the access of the colonized, along lines of class-alliance and class-formation, to the heritage and culture of imperialism. If one assumes an "own-ness" of cultural ground, everything gained through this access was an estrangement. Every attempt at consolidating a cultural ground by this means would definitively exclude the peoples without access to that presumed ground. As I mentioned at the beginning of this paper, the current mood, in the radical fringe of humanistic pedagogy, of uncritical enthusiasm for the Third World makes a demand upon the inhabitant of that Third World to speak up as an authentic ethnic fully representative of his or her tradition. This demand in principle ignores an open secret: that an ethnicity untroubled by the vicissitudes of history and neatly accessible as an object of investigation is a confection to which anthropologists, early colonials, and European scholars partly inspired by the latter, *as well as* indigenous elite nationalists, by way of the culture of imperialism, contributed their labors. (I have hinted at this open secret also in the matter of the construction of "Hinduism"; see note 9.)

There is a great deal to be said about this unexamined negotiation between U.S. Third-Worldism—whether conceived as an alternative to or as an expression of "the Left"—and the construction of the object of colonialism/nationalism, each legitimizing the other. I will draw attention to only one point here: that the current negotiation may be no more and, of course, no less, than a displacement of the negotiation between colonialism and nationalism, even as the latter was ostensibly and, in its context, powerfully taking a stand against the former. As Partha Chatterjee writes:

> The contradictory pulls on nationalistic ideology in its struggle against the dominance of colonialism [led to] . . . a resolution which was built around a separation of the domain of culture into two spheres—the material and the spiritual. It was in the material sphere that the claims of Western civilization were the most powerful. Science, technology, rational forms of economic organization, modern methods of statecraft—these have given the European countries the strength to subjugate non-European peoples and to impose their dominance over the whole world. To overcome this domination, the colonized people must learn those superior techniques of organizing material life and incorporate them within their own cultures. This was one aspect of the nationalist project of rationalizing and reforming the traditional culture of their people. But this could not mean the imitation of the West in every aspect of life, for then the very distinction between the West and East would vanish—the self-identity of national culture would itself be threatened. In fact, as Indian nationalists in the late 19th century argued, not only was it not desirable to imitate the West in anything other than the material aspects of life, it was not even necessary to do so, because in the spiritual domain the East was superior to the West. What was necessary was to cultivate the material techniques of modern Western civilization while retaining and strengthening the distinctive spiritual essence of the national culture. This completed the formulation of the nationalist project, and as an ideological justification for the selective appropriation of Western modernity it continues to hold sway to this day.[35]

Within this scenario, the *Gita* once again comes to occupy an important place in the representation of the spiritual and cultural sphere.[36] It is now declared by the nationalists to have a timeless core that is suprahistorical rather than not-yet-historical, as in Hegel. My concept-metaphor of the graphing of time operates here, however vestigially, by way of the notion of the perennial structures of the universal human mind. I would argue that this notion is a displacement of what we have metonymically named "Hegel," just as nationalism in many ways is a displaced or reversed legitimation of colonialism.

With this brief introduction to the refraction of "Hegel" into the colonial subject, lest we mistake the latter for the figure that occupies the (im)possible perspective of the native informant or the implied contemporary receiver, I proceed to quote now from three sources: first, *Essays on the Gita* (1916), a meditative text by the celebrated nationalist-activist turned sage Aurobindo Ghose; secondly, *The Hindu View of Life* (1927), an authoritative text by nationalist-philosopher turned statesman Sarvepalli Radhakrishnan; and, finally, *Marxism and the Bhagvat Geeta* (1982), a mechanical marxist text that would be held in tolerant contempt by the indigenous sophisticate.[37]

In the first passage, the colonized body politic, the text of the *Gita* itself, has been divided into the material and the spiritual. Its structure has been flattened out. The time-bound material aspect of the text has nothing to do now with the rusing structural liveliness that I have been at pains to point out. Here is Sri Aurobindo:

> No doubt, [in our attempt at reading the *Gita*] we may mix a good deal of error born of our own individuality and of the ideas in which we live, as did greater men before us, but if we steep ourselves in the spirit of this great Scripture and, above all, if we have tried to live in that spirit, we may be sure of finding in it as much real truth as we are capable of receiving as well as the spiritual influence and actual help that, personally, we were intended to derive from it. And that is after all what Scriptures were written to give; the rest is academical disputation or theological dogma. Only those Scriptures, religions, philosophies which can be thus constantly renewed, relived, their stuff of permanent truth constantly reshaped and developed in the inner thought and spiritual experience of a developing humanity, continue to be of living importance to mankind. The rest remain as monuments of the past, but have no actual force or vital impulse for the future. In the *Gita* there is very little that is merely local or temporal and its spirit is so large, profound and universal that even this little can easily be universalised.[38]

It is almost as if the entire graph of the spirit's journey has been sea-changed into a deracialized universalism wherein the cultured colonial nationalist can denegate colonialism. "The spiritual experience of a developing humanity" is neither Hindu nor Hegel but a bit of both.

By 1927, the voice of "academical disputation" itself was carrying the torch of legitimation. Here is Radhakrishnan, writing from the University of Oxford:

> The Hindu method of religious reform helps to bring about a change not in the name but in the content. While we are allowed to retain the same name, we are encouraged to deepen its significance. To take a familiar illustration, the Yahveh of the Pentateuch is a fearsome spirit. . . . The conception of the Holy One who loves mercy rather than sacrifice, who abominates burnt offerings, who reveals himself to those who yearn to know him asserts itself in the writings of Isaiah and Hosea. In the revelation of Jesus we have the conception of God as perfect love. The name "Yahveh" is the common link

> which connects these different developments. When a new cult is accepted by Hinduism, the name is retained though a refinement of the content is effected. To take an example from early Sanskrit literature, it is clear that Kali in her various shapes is a non-Aryan goddess. But she was gradually identified with the supreme Godhead. . . . Similarly Krsna becomes the highest Godhead in the *Bhagavadgita* whatever his past origin may have been.[39]

It is now possible for Radhakrishnan to draw a clear and adequate parallel between the development from pre-Aryan to Aryan and that from Judaism to Christianity. The model of a developing spirit of humanity is aligned, in narrative inspiration, to that very "Hegel" who claimed to Humboldt that modern scholarly findings had removed the grounds for claiming transcendental grandeur for the philosophy of the Hindus, a myth perpetrated by Pythagoras and company (see note 18). The agent of Hinduism is the high colonial/nationalist subject who "refines" the religion into its universalist lineaments. (One might note that "Sanskritization" is, literally, "refinement.") The *Gita* is now the fountainhead of the philosophy of the Aryans, where caste is revealed as the secret of freedom quite in keeping with Marx's famous line "Man makes his own history but not of his own free will" (see note 29).[40]

There is a certain loss of style in the descent or shift from the high culture of territorial imperialism to the vulgarity of indigenous traffic in neocolonialism. Although the ingredients of the earlier universalization of the *Gita* can still be encountered, more typical is a muscular fundamentalism or nativism. Pitted against it is an equally muscular "marxist" idiom that is routinely and perhaps understandably impatient with the folds and pleats of ancient texts. Commenting on the relatively exceptionalist model of the *sthitaprajna,* or the stable-in-knowledge offered by the absolutely exceptional Krishna, such an approach has this to say:

> We have the famous verse which says, "What is night for all creatures is wakefulness for him. What is wakefulness for the creatures is night for him." So, what is light for you and me is darkness for the *sthitaprajna,* what is darkness for us is light for him. The implication is clear. The masses are sunk in ignorance, greed, voluptuousness, temptation, violence, and what not. The one who has seen Light is untouched by all human weaknesses.[41]

None of the philosophical presuppositions of the Hegelian or nationalist fabulations is called into question here. The extraordinary ways in which the text wins assent are necessarily ignored. Indeed, the nationalist admiration for the *Gita,* in the interest of preserving a sense of "national continuity," however spurious, is seen as an ennobling alibi in its time and place.

The impulse toward a new pedagogy of the Third World in the U.S. cannot not articulate itself in the chain of displacements that I put together here in such broad linkings. The least it can do is not undertake to restore the "historical Indian" obliterated by the Hegelian chronotypograph and lurking in the generalized indigenous soul today. There is no historically available authentic (*eigentlich*) Indian point of view that can now step forth (*hervortreten*) and reclaim its rightful place in the narrative of world history. As literary critics and teachers, we might teach our students the way to informed *figurations* of that "lost" perspective. The geopolitical postcolonial situation can then serve as something like a paradigm for the thought of history itself as figuration, figuring something out with "chunks of the real."

At least nationalism and marxism can be seen as positive negotiations (no less benevolent, certainly, than the sanctioned ignorance of General Cultural Studies) with the epistemic graphing of imperialism. For some of the shadow areas in the micrology of the manipulation of law and history, cutting across the body of the great narrative of imperialism, no good word can be said. And our disciplinary goodwill can become complicitous with those areas without the vigilance I attempt to dramatize here. I cannot illuminate those shadow areas in any detail in this broad focus. Let me rather make an impertinent pedagogic move. I will take the liberty of setting this task for aspiring scholars in cultural studies: Figure out the relationship between my account of the Narrative of the Chronotypography of Imperialism and the story told in the following report of the International Commission of Jurists and the Christian Conference of Asia:

> In some countries [of the south and southeast Asian region] the denial [of basic civil and democratic rights to the variously disenfranchised] is built into the constitution and the laws, while in some areas it masquerades under the guise of religious fundamentalism [I have been arguing, with the support of writers such as Chatterjee and Jayawardena, that the two are displacements of each other in the postcolonial discursive formation]. . . . In some cases, the legislations carry the same names—like the Official Secrets Act in Malaysia and India—or similar names (e.g., National Security Act of India; Public Security Act of Nepal; National Security Law of South Korea; Internal Security Act of Malaysia and Singapore; Internal Security Act of Pakistan). . . . Religion, or tyrannical doctrines in the name of religion, are being woven into the constitutions of some countries to suppress [these] rights [of ethnic minorities, women, and so on]."[42]

Look now at the language being used by the indigenous Third World elite to describe the foundations of such practices: "the mono-doctrine of

'Panchshila,' a compulsory state ideology which comprises: 'Belief in the one supreme God, just and civilized humanity, and the unity of [the nation]."

If we remain caught in the shuffle between claims and counterclaims upon a legiferant and adjudicating chronotypography—no disciplinary formation can *fully* avoid it—the only alternative to the hyperbolic admiration for the authentic ethnic might be to proclaim:

> This challenged giant [the United States] . . . may, in fact, be on the point of becoming a David before the growing Goliath of the Third World. I dream that our children will prefer to join this David, with his errors and impasses, armed with our erring and circling about the Idea, the Logos, the Form: in short, the old Judeo-Christian Europe. If it is only an illusion, I like to think it may have a future.[43]

It is in order to take a distance from this reasonable binary opposition that we might be able to make use of the (im)possible perspective of the native informant. The *possibility* of the native informant is, as I have already indicated, inscribed as evidence in the production of the scientific or disciplinary European knowledge of the culture of others: from fieldwork through ethnography into anthropology. That apparently benign subordination of "timing" (the lived) into "Time" (the graph of the Law) cannot of course be retraced to a restorable origin, if origin there is to be found.[44] But the postcolonial reader and teacher can at least (and persistently) attempt to undo that continuing subordination by the figuration of the name—"the native informant"—into a reader's perspective. Are we still condemned to circle around "Idea, Logos, and Form," or can the (ex)orbitant at least be invoked?

The Temporality of Rhetoric

DOMINICK LACAPRA

I WOULD LIKE TO approach the problem of temporality in history and criticism through a discussion of two apparently contrasting interpretations of Romanticism: M. H. Abrams's *Natural Supernaturalism*[1] and Paul de Man's "The Rhetoric of Temporality."[2] De Man's seminal essay appeared two years after Abrams's book, but in many ways it seems to be a condensed critical response to Abrams's magisterial and capacious argument.[3] The two works share a number of features. For example, both treat Romanticism less as a discrete period in literary history than as a movement that both had deep-seated roots in the Western tradition and initiated a modern problematic in literature and criticism. Indeed the two critics tend, whether explicitly or implicitly, to rethink the concept of temporality itself in terms of interacting processes of displacement and (to a limited extent) condensation. Along the way, a notion of periodicity displaces that of periodization, and uneven developments or modes of repetition with variation or change become evident in the history of literature.

But the two critics differ significantly in their understanding of displacement, Abrams seeing it as a nondisruptive modulation or *sanfte Bewegung* in the interest of higher "symbolic" unity and de Man as radical disjunction or decisive difference. Abrams, moreover, locates "true" Romanticism in the "mature" works of the writers he treats (particularly William Wordsworth) and presents "Modernism" (beginning with such figures as Baudelaire) as a "negative" deviation from Romantic "positives." De Man, however, locates the "authentic" voice of Romanticism in "pre-Romantics" such as Rousseau, Schlegel, and E. T. A. Hoffmann—though the voice is at times recaptured in late Romantics such as Baudelaire—and views later attempts at "symbolic" resolution or higher unity as falls from insight into mystified blindness. Indeed while for Abrams Romanticism is at least a qualified success story on an imaginative level, for de Man it is a tale of repeated defeat and aborted apocalypse. The paradoxical allegory of de Man's own reading is that the "authentic" voice of Romanticism tells of the continual frustration of Romantic ideals and dreams, particularly the dream of unity with the desired other. Thus while Abrams seeks to emulate the great Romantic "positives" in his own approach to criticism, for de Man Romanticism is ever and again a campaign waged disastrously out of season—an internally divided movement misconstrued at times by its proponents and typically by its interpreters.

Yet in offering their seemingly antithetical appreciations, Abrams and de Man tend to act out rather than to thematize or work theoretically through the problem of the relation between continuity and discontinuity, sameness and difference, repetition and change (whether gentle or traumatic). With variations in "blindness" and "insight," their texts thus transfer into their own dynamics the very problem in the movement of temporality that is their manifest object of investigation. And each in its own way relies on a very restricted notion of the relation of a text to its pertinent contexts: Abrams in conventional history-of-ideas fashion reduces context to mere background for literature, and de Man depends upon a displaced formalist version of the binary opposition between "internal" literary intertextuality and "external" empirical history or history of taste. Each thus relies on a prevalent preconceived solution to an intricate problem and fails to pose as an explicit theme of investigation the question of how precisely a text comes to terms with various interacting contexts. The result is a confined and at times dubious notion of historicity and an unargued and equally questionable selection and interpretation of certain contexts rather than others. The (somewhat ironic) point of my own discussion will be neither to deny the differences between Abrams and de Man nor to synthesize their perspectives in a higher unity, but rather to suggest both how they supplement one another and how they require still further supplementation in the interest of a more theoretically informed and sociopolitically effective inquiry into the historicity of literature, culture, and society.

Abrams's rhetoric may today seem old-fashioned, and his "high argument" has certainly become familiar. Like Northrop Frye and Frank Kermode among others, Abrams maintains that older theological and philosophical patterns of thought—especially biblical ones—were displaced in the nineteenth century. A primary modality of displacement was of course secularization. Yet the problem is obviously the nature of this displacement of the old by the seemingly new. Abrams tends to take the trauma out of displacement in general and secularization in particular; his harmonizing retrospect lacks the atmosphere of "crisis" that is so evident at least as a rhetorical *topos* in the nineteenth century. Instead he tends to stress continuity with the past and unity in the present among those employing traditional patterns. Abrams does note in passing (p. 183) certain differences in the relation of the Romantics to the past. They were "this-worldly," and they put faith in a version of progress toward higher unity instead of a return to a simple, static origin. But his own decided inclination in interpreting Romanticism as "spilt religion" (in the ironic phrase of T. E. Hulme) is to emphasize saving unity and, by the way, to exclude or downplay irony:

> The Romantic enterprise was an attempt to sustain the inherited cultural order against what to many writers seemed the imminence of chaos; and the resolve to give up what one was convinced one had to give up of the dogmatic understructure of Christianity, yet to save what one could save of its experiential relevance and values, may surely be viewed by the disinterested historian as a display of integrity and of courage. Certainly the greatest Romantic writers, when young and boldly exploratory, earned the right to their views by a hard struggle. (P. 68)

When younger, these writers were often seized by world-shaking, utopian hope inspired by the French Revolution. When older, they became at times a bit disenchanted and darkly disquieting. But in their virile maturity, they presumably saw the light: the Revolution became a figure of hope betrayed, and they turned from sociopolitical radicalism to the hope for personal secular redemption through an apocalypse of the mind and imagination. Leaving the low road of politics, they took the high road of inwardness in seeking a legitimate marriage between self and nature—a symbolic unity whose desirability it is the task of the humanistic critic to recall in his own "high argument" and to defend against threats of chaos such as those posed by an "adversarial" culture beginning toward the end of the nineteenth century.

For Abrams, Wordsworth is a central figure who appears, appropriately enough in the light of Abrams's thesis, at the beginning, the middle, and the end of the book. On one level, the very structure of *Natural Supernaturalism* resonates with the argument about the nature of Romanticism. Why is the Wordsworthian prospect a privileged one? For Abrams, Wordsworth is central because he is an exemplar: both a typical representative of the tendency to displace and secularize traditional patterns and one of the highest and best exponents of this "Romantic" tendency. In Wordsworth the imagination of the individual in direct communication or even communion with nature is the officiant of a legitimate marriage and the means of secular redemption. As Abrams indicates in his discussion of the importance of "inner-light" theologians, it is the more Protestant strand of Christianity that is secularized here. (One might even be tempted to speak of the Protestant ethic and the spirit of Romantic poetry.) Poetry brings inner redemption for the individual alone, for the imagination unites the mind and nature in a secular sacrament that transcends or evades intermediaries (including society and politics), depending on one's point of view. Although Abrams intimates that inner redemption compensated for loss of faith in collective salvation when the French Revolution turned ugly and bloody, he insists that the "major" Romantics, including Wordsworth, defended art for the sake of life. Indeed Wordsworth was preoccupied with the common, everyday life of the ordinary person. His democratic, populist ethos itself was a sign of the times that he translated into his own specific idiom. Unlike later "Modernists" as Abrams sees them, the Romantics did not espouse an elitist and deadly ideal of art for art's sake.

Yet at times Abrams notices things that may jeopardize his insistence upon inner or personal redemption and upon higher symbolic unity. He notes, for example, the sociopolitical implications of Wordsworth's transformed use of language—a point that does not entirely accord with a purely personal or inwardly imaginative (and rather unworldly) interpretation of his poetry:

> In the *Essay* of 1815 [i.e., well after the French Revolution] . . . Wordsworth himself points up the fact that his particular mission as a poet has a social as well as a religio-aesthetic dimension. For if his new poetry of the common man and the commonplace is to create the taste by which it is to be enjoyed, it must utterly reform his readers' characters and sensibilities, which have been permeated with class consciousness and social prejudices. (P. 397)

Yet in a contemporary context, the very unfamiliarity of Wordsworth's approach and the lack of ready-made sensibilities to which it might appeal forced the poet, in Wordsworth's words, to "reconcile himself for a season to few and scattered hearers" (quoted p. 398). Abrams also notes impediments to the quest for higher symbolic unity on a more textual level. On the very first page of chapter 1, he indicates how Wordsworth raised difficulties for his endeavor: "In spite of persistent and anguished effort,

Wordsworth accomplished, in addition to *The Prelude* only Book I of Part I (*Home at Grasmere*), Part II (*The Excursion*), and none of Part III, so that, as Helen Darbishire has remarked, all we have of *The Recluse* is 'a Prelude to the main theme and an Excursion from it'" (p. 19). One tends, however, in Abrams's focus on Wordsworth's presumed intentions and ideals, to lose sight of these contextual and textual limitations on his overall project and their possible implications for interpretation. (Although de Man will exclude certain contextual considerations in his own way and bring into play a less harmonistic but still Protestant strand of displaced Christianity, he will of course place certain textual limitations and considerations in the foreground of his analysis.)

The high Romantic argument that must be preserved centers around what Frank Kermode has called the apocalyptic narrative structure with its relation to speculative dialectics. Narrative provides—in a displaced way—on the level of story and events what speculative dialectics provides on the level of theory and concept. In related but non-identical fashion (which Abrams does not see as problematic), both traditional narrative and speculative dialectics seek a redemptive, revelatory unity, totalization, or closure—a making whole again. The pattern or paradigm is one wherein a circular but progressive journey or circuitous quest seeks an end that recapitulates its beginning on a higher level of insight and development. (Hence the significance of the image of the upward spiral in Hegel and others.) In a secular sense, one has a quest for the redemption of meaning and for a form of justification—a higher identity. In religion, an original state of innocence (Eden) gives way to a fall (original sin) that is overcome through a redemptive act (the coming of a messiah), and one is born again. In narrative, a beginning gives way to a middle whose in's and out's, up's and down's, are made sense of in a concordant ending. In speculative dialectics, identity gives way to difference that is overcome in a higher identity. Wholeness is broken through alienation and suffering that are transcended in a higher, greater wholeness. For Abrams, these patterns that "displace" one another in a circle of mutual confirmation may be figures of desire; but they must be maintained, and their approximation at least in "symbolic" experience of the sort poetry provides is a worthy goal of endeavor.

Abrams steers steadfastly clear of what may be termed lower roads to apocalyptic unity through which religion was displaced and secularized in the course of the nineteenth century. One of the most prominent and powerful was of course nationalism, and it is one legacy of the French Revolution that has continued to affect both social life and thought. The more general point is that for Abrams phenomena such as the French Revolution and industrialization, while periodically named and invoked, remain mere backdrops. The question of the extent to which they were inscribed in texts in symptomatic, critical, or transformative ways is not investigated in any sustained fashion. Instead the revulsion experienced by certain "Romantics" on a biographical level seems to foreclose inquiry into the problem of the Revolution's textual and contextual afterlife and permutations. One gets little sense of how the French Revolution both in fact and in figure was itself "displaced" as a traumatic reference point throughout the nineteenth and even into the twentieth century, particularly but not exclusively in France.

One also gets little sense of the way in which commodity fetishism provided a form of symbolic unity that itself involved a certain reliance on processes of displacement and condensation, for Abrams does not address this crucial dimension of Marx's thought. In his famous analysis in section 4, chapter 1 of *Capital,* Marx is quite explicit in asserting that the "social hieroglyphic" represented by the fetishized commodity evokes an "analogy" with "the mist-enveloped regions of the religious world." One may further argue that the commodity fetish is constituted through a displacement of meaning from the work process in capitalism. Work is reduced to technically instrumental labor power and converted into a commodified exchange value that produces other exchange values. The meaning displaced from "living labor" is condensed in the mystified form of detached symbolic value that is reified in the commodity itself. Thus the commodity becomes a substitute not only for a putative lost totality or essence to be regained on a higher level in a utopian future (the apocalyptic aspect of Marx's thought to which Abrams gives exclusive attention) but also for work that—while never *totally* meaningful—viably engages the problem of meaning. Work in the latter sense suggests a specific alternative to commodity fetishism—an alternative that is not reducible to the apocalyptic paradigm. Its realization requires basic institutional transformation in the economy, society, and culture. (One may note in passing that the critique of commodity fetishism can apply to any form of ideological fetishization or essentialization, including that of pure art or formalism, although it is an open question whether this ideology was affirmed by late-nineteenth-century figures such as Baudelaire in the unproblematic form Abrams believes it was.)

An issue more directly germane to the terms of Abrams's argument as he presents them concerns another "low road" to higher unity—that of sacrifice involving scapegoating and victimization. In sacrifice, one also begins with an innocent or purified "identity," one dismembers or immolates it (in reality or symbolically), and one returns to "identity" on a putatively higher level. Scapegoating and victimization in general involve processes of displacement and condensation: displacement of pollution or guilt from the community onto a victim and simultaneous condensation of anxiety-producing phenomena on the latter in order to expel or exorcise them. In the wake of the French Revolution, Joseph de Maistre (one of the few significant literary figures of the time not mentioned by Abrams) furnished a version of old-time religious patterns that included a providential and sacrificial interpretation of the Revolution itself—one in which scapegoating and victimization were prominent. Maistre's rendition of apocalyptic paradigms (as well as more secular permutations of them) jars rather disruptively with Abrams's construction of the high Romantic argument. Yet such a rendition cannot be simply ostracized or even relegated to the margins of the tradition Abrams treats. It is one significant variant of it. And the larger question it raises is that of whether sacrifice is a ritual analogue (or "displacement") of traditional narrative and speculative dialectics—a question Hegel himself seems to anticipate in his reference to history as a "slaughter bench."

A further set of questions Abrams rather self-consciously does not explore or at least severely underplays is suggested by the multiple and even heterogeneous ways in which the often dominant and displaced apocalyptic patterns are also questioned, contested, ironized, satirized, carnivalized, even ragged—and in which other possibilities are evoked. One possibility is to see time itself (and to articulate narrative as well as theory) in terms of repetition with change, instead of simply reducing complex "displacements" and "condensations" to a "right-angled" apocalyptic model dependent upon a unitary origin, a fall, and a higher-level reunion. An attendant possibility is that difference or otherness is not invariably or uniformly negative and evil (a fall from unity). A difficulty in the apocalyptic paradigm is its fundamentalist identification of all difference or otherness with a fall, an identification that amounts to the equation of unity with the good and alterity with alienation. The very tendency to scapegoat the other through exclusion or immolation is of course facilitated by such an equation. Through scapegoating and victimization, one gets the other back into the fold in one way or another.

Abrams himself is inclined to present all basic questioning of the dominant patterns as exclusively negative—as affronts to a noble dream or a high argument. And he at times gives way to at least methodological scapegoating of tendencies or figures that depart from his argument. All of "modernism" is rather indiscriminately treated as "other" and as negative:

> Salient in our own time is a kind of literary Manichaeism—secular versions of the radical *contemptus mundi* of heretical Christian dualism—whose manifestations in literature extend back through Mallarmé and other French Symbolists to Rimbaud and Baudelaire. A number of our writers and artists have turned away, in revulsion or despair, not only from the culture of Western humanism but from the biological conditions of life itself, and from the life-affirming values. They devote themselves to a new Byzantinism, which T. E. Hulme explicitly opposed to the Romantic celebration of life and admiringly defined as an art which, in its geometrized abstractness, is "entirely independent of vital things," expresses "disgust with the trivial and accidental characteristics of living shapes," and so possesses the supreme virtue of being "anti-vital," "anti-humanistic," and "world-rejecting." Alternatively, the new Manichaeans project a vision of the vileness, or else the blank nothingness of life, and if they celebrate Eros, it is often an Eros *à rebours*—perverse, hence sterile and life-negating. (Pp. 445–46)

Similar arguments have been addressed to recent critical tendencies, prominently including deconstruction. One need not reverse the binaries on which Abrams relies (or even deny that dubious features in an argument emerge when one becomes fixated at a phase of simple reversal of an adversary's position) in order to note that Abrams himself tenders an extremely Manichaean argument and proffers a one-dimensional critique of putatively one-dimensional tendencies. The very homogenization of a heterogeneous set of "adversarial" forces and critical departures from the apocalyptic paradigms itself sets the stage for a scapegoating mechanism.

Even with respect to the early nineteenth century, that is, before the eruption of the "Modernist negatives," Abrams is very selective in his treatment of Romanticism, and his procedure intimates that, if "negatives" there be, they cannot be localized in the later nineteenth century. For example, Abrams excludes Byron from his study, and his reason for so doing also affects his treatment of other figures such as Carlyle and Nietzsche: "Byron I omit altogether; not because I think him a lesser poet than the others but because in his greatest work he speaks with an ironic counter-voice and deliberately opens a satirical perspective on the vatic stance of his Romantic contemporaries" (p. 13). One might think that Abrams here offers an excellent reason for including Byron insofar as even the "genuinely" highest and best should have their ironic and parodic doubles. Abrams in general excludes consideration of forces grouped by Mikhail Bakhtin under the rubric of the carnivalesque, and his resultant understanding of even figures he chooses to discuss is rather one-sided. He seems unable to entertain the possibility that irony and parody may be both critical and reinvigorating—and in ways that need not necessarily imply that one does not respect or take seriously and even value their objects. (The latter depends on the object and the situation.) Abrams himself refers to Carlyle's *Sartor Resartus* in an excellent formulation that would seem to make the same point I am trying to make. He sees the text as "a serious parody of the spiritual autobiography which plays with and undercuts the conventions it nonetheless accepts" (p. 130). (Abrams also notes the humor in Hegel that is evident, for example, in his use of puns.) But from Abrams's own treatment of Carlyle (or Hegel), one sees predominantly the confirmation of the model of a spiritual quest for higher identity. One gets little sense of how *Sartor Resartus* is a tremendously critical, remarkably hybridized, and outrageously funny Menippean satire—a stunning example of carnivalized literature or both high and low comedy of ideas in which the very acceptance of certain conventions is put into "deep" play. Nietzsche himself in *Natural Supernaturalism* is tailored into an unproblematic exponent of the dialectical quest for higher synthesis and the pattern of imaginative redemption through art. He becomes a kind of offbeat and dandified Hegelian, treated in the same terms as Marx—terms that suit entirely neither Nietzsche, Hegel, nor Marx (whatever their own differences may be).

If the carnivalesque tends to be mentioned only in passing and to be excluded from sustained investigation, another significant "displaced" tradition in modern thought is treated in only one of its important guises. Hermeticism is discussed only to the extent to which it, via Neoplatonism, conformed to and confirmed Christianity. One sees well the importance of Ficino and of those with similar perspectives. One gets little idea of why Hermeticism was opposed by the Catholic Church, why Bruno was sent to the stake (a fact not mentioned), or why Hermetic tendencies were attacked by Cartesians and often seen as irrational and suspect in later times. More heterodox forms of Hermeticism, for example in the work of Blake or Nietzsche, go unnoticed.

Abrams, not to put too fine a point on it, is little concerned with the ambivalences, stresses, and strains in the Western tradition—by the internal contestants that can be excluded or confined to its margins and condensed into a homogeneous image of the negative adversary only at the cost of repeating, in however displaced and methodological a form, a scapegoat mechanism. Yet there is a curious if not uncanny way in which *Natural Supernaturalism* threatens to contest or even subvert itself and to indicate however unintentionally a threat to its high argument, for it enacts a kind of textual repetition-compulsion. Abrams repeats the apocalyptic paradigms in an almost obsessive way, in wave upon wave of plangent high seriousness, until the hallowed story he tells becomes almost hollow—eroded and made a bit tedious or even senseless. Formulated in less negative terms, the problem posed by this repetition is that of the relations between the quest for "unity" and the forces—both destructive and possibly enlivening—that challenge it. For the implication of my own critique of Abrams is not that unity in all its forms is mere mystification, inauthenticity, or bad faith but that a crucial problem is the actual and desirable interaction of unity (particularly normative and institutional unity) with its various contestants over time. I would further note in passing that, in his later work, Abrams will reiterate the basic arguments of *Natural Supernaturalism,* and his articulation of them, despite occasional shrillness, will become more nuanced and perhaps more persuasive, in part because of his polemical encounter with deconstructive critiques. It is difficult to locate anything that even seems to be a turn in Abrams's development, but his later return to his earlier arguments supplements the magisterial sweep of *Natural Supernaturalism* with a pointed confrontation with some of its most forceful contestants.

Paul de Man begins his "Rhetoric of Temporality" with an indication of the temporality of rhetoric—a historical insight that might be taken farther than he intimates: "One has to return, in the history of European literature, to the moment when the rhetorical key-terms undergo significant changes and are at the center of important tensions." One such crucial change occurs in the later eighteenth century, when "the word 'symbol' tends to supplant other denominations for figural language, including that of 'allegory'" (p. 173). De Man's argument is that this supplanting marks a fall from insight into blindness both at the time and in contemporary critics such as Abrams, for the role of "allegory," with its complex companion "irony," remains active in certain Romantics and reemerges more forcefully in the later nineteenth century in the work of such figures as Baudelaire. Indeed de Man seems less concerned with the temporality of rhetoric than with the pseudohistoricization of synchronic or at least recurrent patterns such as the interplay of "symbol" and "allegory" themselves. But de Man establishes no simple pattern of rise and fall, progress or regress. In fact he severely criticizes a conventional understanding of history that relies on a simple concept of periodization and seeks some teleological coding of complex processes and structures. His own procedure at times suggests a more problematic "model" of temporality in terms of the relative dominance or submission of recurrently displaced forces and tendencies, but the sense of displacement he insistently stresses equates it with disjunction. The question is whether his explicit statements about temporality and history do justice to his own discursive practice and whether that discursive practice in this essay still contains certain blockages (notably in the form of unquestioned binaries) that excessively restrict an understanding of literary displacements.

Adapting arguments made by Jacques Derrida—arguments that Paul de Man in his later work would accommodate in his own way—one might maintain that a deconstructive strategy requires a double inscription or a dual procedure involving two interacting "phases": the reversal of dominant binary oppositions that inscribe relations of discursive and sociopolitical power, and the general displacement and rearticulation of relations in a manner that resists binarism (as well as its attendant scapegoating) and enacts supplementary relations of mutual marking in a transformed field. The phase of reversal is never entirely transcended, for binaries tend to be reconstituted and new "origins" or "centers" are generated; but it is important not to remain fixated at the phase of reversal since to do so would bring about an inverted form of ideological fetishization or essentialization. In this sense, deconstruction (like the critique of ideology to which it cannot be reduced but that may be seen as its necessary sociopolitical supplement) does not take place once and for all; it is a recurrent activity that remains marked by its "objects," whose recuperative power it nonetheless resists.

"The Rhetoric of Temporality" was of course written before de Man attempted explicitly to reformulate his approach to criticism in terms of deconstruction. But it already puts into play certain deconstructive strategies: reversal on its thematized or theoretically explicit level and elements of general displacement in its "ironic" discursive practice. But the absence of an attempt to articulate the relation between theory and practice may facilitate the tendency to remain fixated at the phase of reversal, seemingly to regenerate a new absolute but inverted "center" or "origin," and to allow certain undeconstructed binaries (notably that between the "inside" and the "outside" of literature) to remain active even on the level of discursive practice. Hence, in this essay at least, it is difficult to know whether one should refer to de Man's rigorous unreliability or to his unregulated equivocation.

With reference to the problem of history, de Man provides cogent and powerful arguments against dubious "traditional" conceptions of temporality, particularly the idea of time as organic continuity or genetic development. But he stops short of a rearticulation of historical understanding, at times reverting to an insistently negative or "nihilating" notion of temporality. His most decisive gestures are to reverse the binary oppositions operative in the thought of such critics as Abrams and to oppose the disjunction and difference operative in allegory and irony to the "symbolic" nature of unity and identity. Yet he appears to be involved in the same quest as Abrams, for he ostensibly seeks the true meaning and authentic voice of Romanticism.

Of "allegorizing tendencies" in Western European literature between 1760 and 1800, de Man states: "Far from being a mannerism inherited from the exterior aspects of the baroque and the rococo, they appear at the most original and profound moments in the works, when an authentic voice becomes audible" (p. 188). The use of terms such as "authentic," "inauthentic," and "bad faith" is more than a mannerism in this essay. Insight seems in fact to be identical with authenticity, and blindness with inauthenticity and bad faith. The terms indicate how an existential rhetoric and conceptualization (particularly in its Sartrean variant) double and displace proto-deconstructive strategies and motifs. De Man's own rhetoric will of course change in this respect, but it is an open question whether his later use of explicitly deconstructive terminology and modes of thematization marks a clear-cut disjunction, a pattern of basic continuity, or a more subtle mode of repetition and change wherein the traumatic breaks with "The Rhetoric of Temporality" are themselves uneven, heterogeneous, and often difficult to trace.[4] The obvious questions agitating this early seminal essay are: What is authentic insight? What is inauthentic blindness or bad faith? How do they relate to one another and to the problem of temporality or historicity? And what is one to make of the seemingly symbolic relation conveyed in the equations of insight with authenticity and blindness with inauthenticity, as well as in other dimensions of de Man's own argument? For de Man often employs the language of the symbolic to argue for the disjunctive and allegorical.

Issues are further complicated by the manner in which traces of formalism double and displace an existential thematic. De Man often relies on a binary opposition between what is intrinsic and what is extrinsic to literary texts, most blatantly when he peremptorily dismisses Daniel Mornet's interest in the role of the *jardin anglais* in Rousseau's *La Nouvelle Héloïse* as merely superficial history of taste and insists upon the role of literary "intertexts"—as if the significance of the latter simply excluded the pertinence of the former. Moreover, the undeconstructed "inside/

outside" binary induces a tendency to conflate the distinction between fiction and nonfiction with the opposition between literature or art and the actual, factual, or sociohistorical world. For example, in discussing Baudelaire and Schlegel, de Man asserts: "Far from being a return to the world, irony, at the second power as 'irony of irony' that all true irony at once has to engender asserts and maintains its fictional character by stating the continued impossibility of reconciling the world of fiction with the actual world" (p. 200). "True" irony is here equated with de Man's own understanding of Romantic irony, especially as infinitely regressive self-reflexivity or linguistic self-referentiality and *mise en abîme*. Yet in seemingly contradictory fashion de Man also asserts: "Irony comes closer [than allegory] to the pattern of factual experience and recaptures some of the factitiousness of human existence as a succession of isolated moments lived by a divided self" (p. 207). In this last quotation, de Man may mean "facticity" instead of "factitiousness," but the slip (if it is a slip) in the direction of sham and artifice brings out the problematic nature of any concept of a "world of fiction" inhabited by literary texts or art works cut off from the "actual world." For this "two-world" theory is a limiting form of binarism (also manifest in the implicit assumption that the only alternative to reconciliation is disjunction). It reifies or ideologically essentializes what is at best a hyperbolic understanding of a modern dilemma—one wherein certain possibilities seem excluded from social life only to be relegated to literature and art construed as a separate realm. Such a theory obscures the processes by which a separatist institution of literature is elaborated as well as the functions it serves, and it simultaneously conceals the role of "fiction" in "actual life." More generally, it provides no critical, nonreductive basis on which to raise the question of the actual and desirable interaction between literature or art and social life. Rather it facilitates an ideological conception of the status of the literary text that may (mystifyingly) see itself as the demystification of ideology—a conception that is a displaced, perhaps abortive form of transcendental metaphysics in the guise of pure figurality or fiction. The result in criticism is to generate a seemingly impenetrable barrier between texts and contexts that, insofar as they are not literary or linguistic in a formal sense, are either ignored or deemed "exotopic."

On the level of fixated reversal that generates a new metaphysical center or origin, authentic insight—if one can speak about it at all without immediately lapsing into inauthenticity (which is doubtful)—seems to be an unlivably blank apprehension of absence, pure difference, or nothingness, a "look" into the abyss of radical disjunction. With respect to this blinding insight, the only language that resists inauthenticity or at least asserts the knowledge of inauthenticity (without being itself authentic) is figuration that has an ironic awareness of its own figural or fictional status as radically disjoined from phenomenal or empirical reality. This appears to be de Man's ultrasophisticated version of the imagination—the neo-Kantian *Einbildungskraft*. It becomes figurality or fictionality that has an ironic awareness of itself and its uncompromising otherness, indeed its problematic transcendence, especially through endless self-referentiality with its negative thrusts and impasses. Inauthentic blindness or bad faith, by contrast, seems to be a "relapse" into modes of "symbolic" unity between language and world, including both realistic representation and putative marriages of the mind and nature. This relapse itself presumably engenders a false apprehension of (synchronic) figuration—the (blind) play of tropes—that is identified with (literary) history. The "fall" would thus be a fall not from but into unity. And (pseudo)history would be misconstrued figuration. Authentic temporality, by contrast, would itself seem to be the marking and remarking of pure absence, difference, or disjunction. As a result, history would be the history of error or erring, and an impossible "authentic" historicity would seem to engender nothing except the evacuation of signification in a compulsive reiteration of aporias having "no exit":

> The dialectical relationship between subject and object is no longer the central statement of romantic thought, but this dialectic is now located entirely in the temporal relationships that exist within a system of allegorical signs. It becomes a conflict between a conception of the self seen in its authentically temporal predicament and a defensive strategy that tries to hide from this negative self-knowledge. On the level of language the asserted superiority of the symbol over allegory, so frequent during the nineteenth century, is one of the forms taken by this tenacious self-mystification. (P. 191)

> Irony divides the flow of temporal experience into a past that is pure mystification and a future that remains harassed forever by a relapse within the inauthentic. It can know this inauthenticity but can never overcome it. It can only restate and repeat it on an increasingly conscious level, but it remains endlessly caught in the impossibility of making this knowledge applicable to the empirical world. It dissolves in the narrowing spiral of a linguistic sign that becomes more and more remote from its meaning, and it can find no escape from this spiral. The temporal void that it reveals is the same void we encountered when we found allegory always implying an unreachable anteriority. Allegory and irony are thus linked in their common discovery of a truly temporal predicament. (P. 203)

Earlier in the text de Man explicates Baudelaire, but his questionable exegesis seems clearly to merge his own voice with that of his putative object of analysis. The new center is approximated to violence and incipient madness, which are conceived not only as recurrent and variable possibilities (threats and temptations) but as "inherent" and "necessary" components of the seemingly authentic human condition, that is, as essentialized and foundational features of a fixated synchronicity that itself seems coeval if not identical with the "truly temporal predicament":

> Irony is unrelieved *vertige*, dizziness to the point of madness. Sanity can exist only because we are willing to function within the conventions of duplicity and dissimulation, just as social language dissimulates the inherent violence of the actual relationships between human beings. Once this mask is shown to be a mask, the authentic being underneath appears necessarily as on the verge of madness. (P. 198)

Indeed fixated synchronicity seems tantamount to acting out a repetition compulsion, and, as seemingly authentic temporality, it is recurrently contrasted with the mystified phantasm of symbolic unity and totalization. De Man's thought often seems suspended between impossible totalization and a repetition compulsion, especially in terms of a mechanistic play of tropes. One need not agree with M. H. Abrams to have a critical response to this entire problematic. The obvious question one may raise is whether something is repressed or disavowed in the opposition between totality and the repetition compulsion—what Freud referred to as "working through" and Marx as critique. One may also suggest that de Man's essay, insofar as it is protodeconstructive and marked by existential and formalist motifs, approximates to certain seemingly postdeconstructive endeavors. Critics alienated by explicit forms of deconstructive criticism may well try to amalgamate certain of its (unnamed) features with earlier existential and formalist proclivities. In numerous and overdetermined ways, therefore, "The Rhetoric of Temporality" is far from dead in contemporary criticism. For many, it may seem a very live option.

I now turn to three important sections of "The Rhetoric of Temporality": its discussions of Rousseau, Wordsworth, and Baudelaire. This sequence of figures may in certain respects reflect de Man's understanding of Romanticism in the late eighteenth and in the nineteenth centuries. After the publication of "The Rhetoric of Temporality," Rousseau became a major reference point for de Man—a fact that gives his treatment of Rousseau a special significance. De Man argues that, given its role in the critical literature, "there is certainly no better reference to be found than *La Nouvelle Héloïse* for putting to the test the nearly unanimous conviction that the origins of romanticism coincide with the beginnings of a predominantly symbolical diction" (p. 184). De Man does find such diction in the Meillerie episode in the fourth part of the novel. But he notes that the Meillerie landscape as a wilderness is in the terms of the novel emblematic of error. By contrast the garden functions as "the landscape representative of the 'beautiful soul'" (p. 185). And the garden is treated allegorically not only in its function as an emblem of virtue within the novel but in its literary intertextuality as well:

> From the beginning we are told that the natural aspect of the site is in fact the result of extreme artifice, that in this bower of bliss, contrary to the tradition of the *topos,* we are entirely in the realm of art and not that of nature. "Il est vrai," Rousseau has Julie say, "que la nature a tout fait [dans ce jardin] mais sous ma direction, et il n'y a rien là que je n'aie ordonné." The statement should at least alert us to the literary sources of the passage which Mornet, preoccupied as he was with the outward history of taste, was led to neglect. (P. 186)

Before turning to these literary sources, one may note two features of this passage. First, Rousseau relies on an explicit oxymoron in having Julie say that nature did everything in the garden but under her direction. He thereby marks a disjunction in a non-exclusionary way that would seem to intimate the complex role of both nature and human direction or artifice. Affected perhaps by the *vertige de l'hyperbole* that de Man, following Baudelaire, tells us is characteristic of irony, de Man provides an explicit gloss that reductively converts the oxymoron into an either/or choice by maintaining that "the natural aspect of the site is in fact the result of extreme artifice" and that "we are entirely in the realm of art and not that of nature"—hardly what the passage says. Furthermore, his reliance on the opposition between internal literary relations and "the outward history of taste" again implies an either/or decision. What is quite explicit in the exclusion of the seemingly "outward" in this passage is typical of de Man's treatment of contexts in general. He will refer only to "intraliterary" or "intertextual" relations (in the ordinary sense), and he will not even mention the contexts Abrams at least employs as background. In this sense, his reading is formalistic, and he does not attempt to rethink the entire relation between texts and contexts in a manner that would be neither formalistic nor contextually reductionistic and that might therefore question the binary opposition between inner literary history and the "outward history of taste." Indeed de Man here sounds very much like Northrop Frye, and his procedure indicates how even allegory in his sense remains symbolic, that is, separated or cut off ("castrated" in the Freudian sense)—cut off from larger modes of interaction and confined within a specifically literary "realm" or "world."

To make these observations is in no sense to exclude the pertinence of the "literary" aspects of intertextuality that de Man brilliantly locates, but it is to question the framework within which he sets them and certain of the conclusions he draws from his analysis. Indeed the combination of "literary sources" he offers for the garden passage does come as quite a shock, as does the nature of his formulations in articulating their relations to *La Nouvelle Héloïse*. The two sources are Defoe's *Robinson Crusoe* and Guillaume de Lorris's *Roman de la rose*. We are told that "there is hardly a detail of Rousseau's description that does not find its counterpart in the medieval text" (p. 186). Indeed so commanding is the role of this "source" that "Rousseau does not even pretend to be observing. The language is purely figural, not based on perception" (p. 187). Here as frequently in the essay, the cogent point that language or any system of signs cannot be reduced to perception is converted into a binary opposition between language and perception (or empirical reality in general), with the result that a "realm" of language as pure figurality or fictionality is generated.

There is, however, a difference between *La Nouvelle Héloïse* and the *Roman de la rose*, and it indicates the role of *Robinson Crusoe* as a second source for the novel. One has an "exalted treatment of erotic themes" in the *Roman de la rose*. "In *La Nouvelle Héloïse* the emphasis on an ethic of renunciation conveys a moral climate that differs entirely from the moralizing sections of the medieval romance" (p. 187). Thus the detailed "counterparts" accompany "a moral climate that differs entirely" from one text to the other. One may note the rhetoric of "all or nothing" in which one has a starkly paradoxical tendency to qualify an absolute assertion made in the diction of the "symbolic" (i.e., with reference to counterparts or correspondences) with an equally absolute assertion made in the diction of allegorical-ironic disjunction. The possibility that a certain religious heritage or bundle of tendencies within Christianity may be displaced in de Man's own rhetoric is suggested by what he responds to in Rousseau's own response to Defoe. This displacement would provide a context of sorts for de Man's reliance on a problematic focused on the "temptation" to "fall" into symbolic unity or meaning—a "temptation" that must be resisted or renounced even if one inevitably "relapses" into an inauthentic mode. Indeed the relatively undisplaced reliance on religious or theological terminology in certain passages of "The Rhetoric of Temporality" is striking. For example:

> The temptation at once arises for the ironic subject to construe its function as one of assistance to the original self and to act as if it existed for the sake of this world-bound person. This results in an immediate degradation to an intersubjective level, away from the "*comique absolu*" into what Baudelaire calls "*comique significatif*," into a betrayal of the ironic mode. Instead, the ironic subject at once has to ironize its own predicament and observe in turn, with the detachment and disinterestedness that Baudelaire demands of this kind of spectator, the temptation to which it is about to succumb. It does so precisely by avoiding the return to the world . . . , by reasserting the purely fictional nature of its own universe and by carefully maintaining the radical difference that separates fiction from the world of empirical reality. (P. 199)

In contrast to de Man's later "technical" language, this allegory of "ironic" transcendence and worldly temptation gives one a vivid sense of how we still may be fighting, in displaced and more or less symbolic ways, the wars of religion. One also gets a feeling for the manner in which a critic such as Abrams may appear as excessively compromising or Erasmian when viewed from a more "rigorous" if not intransigent point of view.

To return to de Man's treatment of Rousseau (have we ever left it?), it may be noted that de Man stresses that recent interpretations "have reversed the trend to see in Defoe one of the inventors of a modern 'realistic' idiom and have rediscovered the importance of the puritanical religious element to which Rousseau responded" (p. 187). Indeed "the same stress on hardship, toil, and virtue is present in Julie's garden, relating the scene closely to the Protestant allegorical tradition of which the English version, culminating in Bunyan, reached Rousseau through a variety of sources, including Defoe. The stylistic likeness of the sources supersedes all further differences between them" (p. 188). De Man does not of course present Rousseau simply as a secularized Puritan, and we cannot see de Man simply as a secularized sectarian Protestant or Jansenist despite the seemingly obvious relations between "Augustinian" tendencies in Christianity and de Man's ascetic "rigor," stark sense of paradox, ironic (yet somehow Bunyanesque) allegory of temptation and renunciation, and iconoclastic notion of the disjunctive relation between sign and meaning. One may, however, observe that the tendency to assert the priority of allegory in the diction of symbol ("the stylistic likeness of the sources that supersedes all further differences between them") becomes paradoxically pronounced in de Man's ironically "self-deconstructive" conclusion that *La Nouvelle Héloïse* achieves ("symbolic") resolution through renunciation and closure through the choice of allegory:

> This conflict [between the erotic and the puritanical] is ultimately resolved in the triumph of a controlled and lucid renunciation of the values associated with the cult of the moment, and this renunciation establishes the priority of an allegorical over a symbolic diction. The novel could not exist without the simultaneous presence of both metaphysical modes, nor could it reach its conclusion without the implied choice in favor of allegory over symbol. (P. 188)

Yet, to paraphrase Derrida, would it not appear that we are here in a region—let us call it provisionally a region of historicity—in which the category of choice seems particularly superficial?

De Man notes a variety of differences between allegory and irony themselves, notably in the extended, diachronic, narrativized, or displaced temporal structure of the former and the pointed, synchronic, aphoristic, condensed temporality of the latter. Yet he also insists:

> The knowledge derived from both modes is essentially the same; Hölderlin's or Wordsworth's [allegorical] wisdom could be stated ironically, and the rapidity of Schlegel or Baudelaire could be preserved in terms of general wisdom. Both modes are fully de-mystified when they remain within the realm of their respective languages but are totally vulnerable to renewed blindness as soon as they leave it for the empirical world. Both are determined by an authentic experience of temporality which, seen from the point of view of the self engaged in the world, is a negative one. The dialectical play between the two modes, as well as their common interplay with mystified forms of language (such as symbolic or mimetic representation), which it is not in their power to eradicate, make up what is called literary history. (P. 207)

Nonetheless, the very interaction between symbolic unity and allegorical-ironic disjunction is played out in the discussion of allegory and irony itself, and the way it unfolds induces one to question the clear-cut delimitation of realms of language as well as the identification of symbol as the sign of mystifying inauthenticity and disjunction as the sign of insightful authenticity. De Man concludes his essay with a discussion of Baudelaire's *De l'essence du rire* as a starting point for reflection about irony, an exegesis of one of Wordsworth's "Lucy" poems as a demonstration of the role of allegory, and a brief comment about Stendhal's *La Chartreuse de Parme* (a self-conscious, extended ironic narrative) as the seeming exception to the rule concerning the aphoristic brevity of irony and the narrativized "duration" of allegory. I would like to indicate that what seems to be the exception is not only relevant to the rule but may be the rule itself insofar as irony and allegory, as well as symbol and disjunction, interact in specific texts or uses of language. This point is indeed seemingly intimated in de Man's ironic choice of an eight-line poem to serve as the exemplar of allegory and of a fairly substantial essay to motivate the discussion of irony.

One may also note the role of a more complex mode of "temporality" that cannot be decisively reduced to the binary opposition between the synchronic and the diachronic or its multiple analogues—a mode of temporality involving complex relations between repetition and change over time. De Man seems close to this notion of repetitive temporality both in his discussion of allegory ("the meaning constituted by the allegorical sign can then consist only in the *repetition* in the Kierkegaardian sense of a previous sign with which it can never coincide, since it is of the essence of this previous sign to be pure anteriority," p. 190) and in his discussion of irony ("irony is not temporary—*vorlaüfig*—but repetitive, the recurrence of a self-escalating act of consciousness," p. 202). But his formulations remain indentured to tendencies such as binarism that I have already touched upon, and he does not explicitly bring them to bear on the interplay of symbol, allegory, and irony in texts. Indeed his passing reference to literary history as constituted by the "dialectical play" of allegory and irony as well as "their common interplay with mystified forms of language (such as symbolic or mimetic representation)," while richly suggestive, remains somewhat abstract, mechanical, and tendentious in formulation. (To generalize rashly, one might say that in his later work de Man would often see this "interplay" in terms of his specific variant of deconstruction wherein elements such as allegory and symbol or metonymy and metaphor undercut each other in aporetic fashion.) More importantly, he does not here (and I think never more than very allusively manages to) relate the notion of repetitive temporality as "displacement" involving more or less disjunctive or traumatic change to the attempt to rearticulate the manner in which texts inscribe various contexts within a larger network or "general text." This effort, while not free of its own "mystifying" temptations (particularly when it takes itself as foundational), may nonetheless suggest a notion of historicity that offers some perspective on the interaction of texts with both "literary" and other contexts of writing, reception, and critical reading—including contexts in the so-called empirical world of "engaged" activity.

De Man turns to the poet who served as the exemplar of the symbolic in Abrams in order to approach the question of whether there can be what Baudelaire terms *la poésie pure* in the form of pure allegory that is truly meta-ironical in that it has "transcended irony without falling into the myth of an organic totality or bypassing the temporality of all language" (p. 204). He quotes this "Lucy" poem of Wordsworth:

> A slumber did my spirit seal;
> I had no human fears:
> She seemed a thing that could not feel
> The touch of earthly years.
>
> No motion has she now, no force;
> She neither hears nor sees;
> Rolled round in earth's diurnal course,
> With rocks, and stones, and trees.

De Man asserts that "the text is clearly not ironic, either in its tonality or in its meaning" (p. 205). His genial interpretation presents it as narrating a movement from blindness or mystification belonging to the past in the first stanza to insight or demystification situated in the "now" of the poem in the second stanza. The event that separates the two states is the radical discontinuity of a death that remains impersonal: its moment of occurrence—an "actual now"—is hidden in the blank space separating the two stanzas. The poem thus narrates a series of now-points: the first past and mystified; the second actual and deadly; the third an "eternal" or "ideal 'now,' the duration of an acquired wisdom," which is engendered by the language of the poem but "is not possible within the actual temporality of experience." Moreover, de Man claims that "the fundamental structure of allegory reappears here in the tendency of the language toward narrative, the spreading out along the axis of an imaginary time in order to give duration to what is, in fact, simultaneous within the subject" (p. 206).

In a manner reminiscent of Heidegger in "The Origin of the Work of Art," de Man notes an "ambiguity" in lines 3 and 4 concentrated in the word "thing." Within the mystified world of the past, the word "thing" might have been used "innocently" or perhaps even in a "playfully amorous way." But the innocuous statement becomes literally true in the retrospective of the eternal "now" of the second part, for "she now has become a *thing* in the full sense of the word, not unlike Baudelaire's falling man who became a thing in the grip of gravity." Indeed "the light-hearted compliment has turned into a grim awareness of the de-mystifying power of death, which makes all the past appear as a flight into the inauthenticity of a forgetting" (p. 205). When read "within the perspective of the entire poem," lines 3 and 4 are ironic, "though they are not ironic in themselves or within the context of the first stanza" (pp. 205–6). Irony here thus seems to depend entirely on a sequential temporal structure that is nonetheless positioned as an illusory effect of language.

De Man's reading, particularly within the context of his essay, is extremely plausible and even compelling. I would, however, like to suggest the possibility of an alternative reading that brings out how a repetitive temporality is enacted in the poem and how a less-contained interaction of irony and allegory may be active in it. First one may note that an ironic or parodic reading of the entire poem is always possible. Indeed it may impishly be prompted by the ostensible role of sustained high seriousness in it. The reading may be effected simply by the recitation of the poem in a sing-song intonation, which its versification almost seems to invite. (A similar ironic or parodic reaction may also be stimulated by certain phrases in de Man's exegesis that border—whether intentionally or not—on purplish pop-existentialism, for example, "an eternal insight into the rocky barrenness of the human predicament prevails," p. 206.) Moreover, de Man's hypothesis concerning the role of a sweet young thing in the first stanza is of course fanciful. One might argue that the ambiguity of lines 3 and 4 is much more extensive and that death is already marking life in the first stanza, where these lines could be read quite literally to mean that "she" seemed insensitive to aging as it did its inevitable work. The blank space would thus not simply separate the two stanzas in a before-and-after sequence, but would itself be at work in both of them as well—repeated with a significant, indeed a traumatic, variation. Moreover, rather than being simply sequential in terms of now-points, the stanzas could be seen as related by the chiasmus of an ironic reversal that would crisscross and contest the allegory. The chiasmus would apply most clearly to the interaction of life and death, for if death already marks life in the first stanza, there is a hint of life emerging from death in the second. The last line is especially significant in this respect. Rocks and stones seem redundant as mineral forms of anti-life, but with trees one returns—however faintly and with a "shock of mild surprise"—to life and the promise of renewal. Thus the poem as a whole could be read as bearing witness in a muted, memorialized, and relatively decontextualized manner to a repetitive rhythm of life and death—a fully ambivalent temporality—that in other contexts might be institutionalized in a social structure, like carnival, with a clear bearing on the actual temporality of experience. This reading would be more in keeping with a Freudian interpretation of the nature of *Nachträglichkeit* or deferred action, toward which de Man himself seems at times to be gesturing in his exegesis.

De Man begins his extensive discussion of Baudelaire's *De l'essence du rire* with a point that is in one sense well taken and in another specious. He tries to justify or motivate the difference between his treatments of allegory and irony by arguing that a "historical de-mystification" was necessary in the case of allegory since its "mystification is a fact of history and must therefore be dealt with in a historical manner before actual theorization can start." By contrast:

> In the case of irony one has to start from the structure of the trope itself, taking one's cue from texts that are de-mystified and, to a large extent, themselves ironical. For that matter, the target of their irony is very often the claim to speak about human matters as if they were facts of history. It is a historical fact that irony becomes increasingly conscious of itself in the course of demonstrating the impossibility of our being historical. In speaking of irony we are dealing not with the history of an error but with a problem that exists within the self. (P. 194)

The later de Man would no doubt substitute "language" for "the self." But even with this substitution, his argument would remain problematic. In one sense, he might be read to mean that the very structure of historicity is in some pertinent sense ironic and thus it makes no sense to try to write a standard history of irony that would, for example, seek some oriented development or delimit well-defined periods in the unfolding of irony. But this reading would imply the impossibility of our not being historical or engaged by the problem of irony. On another reading, however, de Man himself relies on the standard and dubious binary opposition between history and structure. He seems to occlude the possibility of significant variations in the recurrent role of irony over time whose different figurations and implications for social life could be traced. The occlusion affects his own discourse and his interpretation of Baudelaire's essay, for de Man himself seems symptomatically to underwrite certain modern tendencies by extracting irony from its likely involvement in a larger network of carnivalesque forces and construing it in rather narrow, intellectualistic terms. It seems able to provoke a rictus but never a belly laugh. This approach may itself be a hyperbolic and potentially unsettling testimony to modern problems—particularly insofar as de Man stresses the relation between irony, seemingly open-ended freedom, and lucid madness. But it provides little critical perspective on these problems and no indication of their possible transformation—and I mean transformation not into an impossible "symbolic" totality or into a state definitively transcending all radically disorienting threats and temptations but into a significantly different network of relations, including the relation between work and "carnivalesque" play.

De Man begins his discussion of *De l'essence du rire* by noting the role of the fall in Baudelaire, specifically the simple but pointed spectacle of a man falling in the street. He quotes Baudelaire as writing "Ce n'est point l'homme qui tombe qui rit de sa propre chute, à moins qu'il ne soit un philosophe, un homme qui ait acquis, par habitude, la force de se dédoubler rapidement et d'assister comme spectateur désintéressé aux phénomènes de son *moi*" ("It is not the man who falls who laughs at his own fall, unless he is a philosopher, a man who has acquired by habit the power to double himself rapidly and to witness as a disinterested spectator phenomena involving his own ego," my translation). De Man observes that "the accent falls on the notion of *dédoublement* as the characteristic that sets apart a reflective activity, such as that of the philosopher, from the activity of the ordinary self caught in everyday concerns." He goes on to remark further that it is in seeming asides such as this that Baudelaire introduces "the notion of self-duplication or self-multiplication" that "emerges at the end of the essay as the key concept of the article, the concept for the sake of which the essay has in fact been written" (p. 194).

De Man is of course to the point in bringing out the importance of the phenomenon of *dédoublement* in Baudelaire's understanding of "irony," although one might contend that this phenomenon attests not to the role of disjunction alone but to the intricate and shifting interplay of disjunction and conjunction in language and the self. Two features of de Man's introductory comments are puzzling, however. In presenting *dédoublement* as the "key concept" of the article, he offers a teleological interpretation, which is precisely what irony was said to jeopardize. Moreover, he fails to note that when the concept reemerges at the end of Baudelaire's article, it is with a significant variation, for there it is linked not to the philosopher but to the artist and to artistic phenomena. All artistic phenomena are said to "denote in the human being the existence of a permanent duality, the power to be at the same time oneself and another." And "the artist is not an artist unless he is a double being and is unaware of no phenomenon in his double nature."[5] This displacement or doubling is significant, for it intimates that the artist may be seen as a displacement of the philosopher (a theme de Man himself takes up in his later work). In addition, the philosopher, while a "rare" case in life, is not entirely separated from "ordinary reality." He is in fact typified as someone who reacts in a certain self-critical way to one of the most ordinary events in life: falling in the street. Nor, as we shall see, is Baudelaire's idea of art for art's sake totally removed from the world.

A second point is that de Man mentions only in passing a second sense of falling that indicates how falling is itself doubled in structure—the religious sense. This slight is difficult to see as unintentional since the religious fall is crucial to Baudelaire's entire argument. Laughter for him is postlapsarian, and the satanic mark of the fall touches all laughter in one way or another. Yet falling, while retaining its religious aura, is shifted in the direction of a repetitive process rather than inserted into an apocalyptic paradigm.

It is the height of irony that a reader of de Man's essay would never realize that Baudelaire's *De l'essence du rire* is not about irony. In fact the word is never mentioned. This is not to say that the problem of irony is not pertinent to its argument or that it may in its silence say more about irony than many full-length treatises devoted to the elusive topic. But it is noteworthy that Baudelaire discusses caricature, the grotesque, the comic, and laughter—that is to say, a larger "family" or congeries of forces to which irony in some sense belongs. It is de Man who isolates irony, and he does so without giving the reader an idea of the problematic nature of his analysis or the larger context in which problems are treated in Baudelaire.

Here it is interesting to notice the way Baudelaire presents his argument—his own mise-en-scène, so to speak. Baudelaire begins by stating that he does not want to write a treatise on caricature but to share with the reader certain reflections that have become a sort of obsession with him. He wants to "relieve" himself ("me soulager") and give himself a better digestion ("en rendre ainsi la digestion plus facile," p. 370). He then employs figures of eating and defecation with respect to a psychological or even psychopathological state. A symptom is to be treated through writing, which is itself figured in manifestly carnivalesque terms. Baudelaire soon makes a further connection between the carnivalesque and the hermetic as he evokes, in terms reminiscent of his procedure in *Les Fleurs du mal,* an alchemical extraction of beauty from ugliness and evil—a spectacle that excites "an immortal and incorrigible hilarity" in man. He adds, "Here is therefore the true subject of the article" (p. 371). It is significant that at least in a brief and allusive way Baudelaire situates his article with respect to two important underground traditions in the modern period—the carnivalesque and the hermetic—and he links the two through hilarity. He thus indicates some awareness of the problem of the displaced recurrence of what has been repressed or suppressed in history.

At the very outset, the repressed reappears in more or less distorted or masked form in both the psychopathological symptom and the work of art—a reappearance that poses the problem of the relation of psychopathology and art to each another and to larger historical sociopolitical processes. Here one can see how a certain set of problems is inscribed not simply in a symptomatic fashion in this text but also in a manner that has important critical and possibly even transformative possibilities. We shall note other ways in which this is the case.

De Man devotes much attention to the contrast between the ordinary, social, or "significative" comic and the absolute comic. He identifies the latter with irony although Baudelaire correlates it with the grotesque. And he treats the two largely as constitutive of a pure binary opposition, with E. T. A. Hoffmann serving as the exemplar of the absolute comic. "Baudelaire insists that irony, as '*comique absolu,*' is an infinitely higher form of comedy than is the intersubjective kind he finds so frequently among the French; hence his preference for Italian *commedia dell'arte,* English pantomimes, or the tales of E. T. H. [*sic*] Hoffmann" (p. 195).[6] Moreover, Baudelaire "rightly considers" Hoffmann "to be an instance of absolute irony" (pp. 198–99). De Man also sees Baudelaire, particularly in his understanding of Hoffmann, as converging with Friedrich Schlegel in an apprehension of irony as "*folie lucide,*" "unrelieved *vertige,* dizziness to the point of madness" (p. 198). In discussing Hoffmann's *Princess Brambilla,* he notes that the drawings of Callot, which are identified as a "source" in Hoffmann's tale, themselves "represent figures from the *commedia dell'arte* floating against a background that is precisely *not* the world, adrift in an empty sky" (p. 200).[7]

One may, however, observe that Baudelaire's argument concerning the comic has a ternary, not a binary, structure and that his reading of its terms, as well as of the role of E. T. A. Hoffmann as an exemplary figure, is quite different from, not to say contrary to, de Man's. The three forms Baudelaire refers to are the definitive absolute, the absolute comic, and the significative comic. The definitive absolute—the highest mode—is not discussed by de Man. It transcends laughter and constitutes an impossible ideal or limit: "From the point of view of the definitive absolute, there is nothing but joy" (p. 375). Joy, moreover, is earlier discussed in the text in clearly symbolic terms: "Joy is *one*. Laughter is the expression of a double or contradictory sentiment; and it is for that reason that it is convulsive" (p. 374). The definitive absolute and joy are discussed in terms that relate them to other figures or issues in the text: the contemplative sage who laughs only when trembling and whose wisdom approaches a higher innocence; the incarnate Word, Christ, who never laughed; and *la poésie pure,* which is in Baudelaire approximated to the symbolic (and not to the allegoric as in de Man). In "pure poetry, limpid and profound like nature, laughter would be lacking as in the soul of the sage" (p. 373). But all of these "corresponding" limiting cases are inaccessible to fallen man. The absolute comic itself "can be absolute only relative to a fallen humanity" (p. 375). It is nonetheless closer to unity than is the significative or social comic: "The element [of the significative comic] is visibly double: art and the moral idea; but the absolute comic, approximating nature much more, presents a unified species [*une espèce* une] and wants to be apprehended by an intuition" (p. 375). The absolute comic is less reflective than the significative, and it erupts suddenly rather than after the fact; in this as in other ways it is closer to the unconscious: "One of the very particular signs of the absolute comic is to be unaware of itself" (p. 378).

It is significant that in Baudelaire the absolute comic is not "infinitely higher" than the significative or social comic. It is certainly seen as superior, but the relation between the two is not constitutive of a pure binary. There is even a curious sense in which it may be lower, for it seems to posit a superiority of man over the rest of nature (in contrast to the significative comic, which rests upon an invidious distinction between man and man). Yet its seeming superiority, which like that of the significative is a derivative of the fall, may be mistaken. In any case, the absolute and significative comic are explicitly treated by Baudelaire in the "impure" terms of supplementary differences and similarities, and they can be active in the same texts, notably those of E. T. A. Hoffmann. Baudelaire states:

> The [significative] comic is, from the artistic point of view, an imitation; the grotesque, a creation. The [significative] comic is an imitation mixed with a certain creative faculty, that is to say, an artistic ideality. Human pride, which always gets the upper hand, also becomes the natural cause of laughter in the case of the grotesque, which is a creation mixed with a certain imitative degree of elements preexisting in nature. I mean that in the latter case laughter is the expression of an idea of superiority not of man over man but of man over nature. One must not find this idea too subtle; that would not be a sufficient reason to reject it. It is a question of finding another plausible explanation. If this one seems farfetched and a little difficult to accept, it is because laughter caused by the grotesque has something more profound, axiomatic, and primitive, which is very close to innocent life and absolute joy, than is the laughter caused by the comedy of manners. There is, between these two laughters, abstraction made of the question of utility, the same difference as between the engaged [*intéressée*] literary school and the school of

> art for art's sake [*l'art pour l'art*]. Thus the grotesque dominates the comic with a proportional elevation. (P. 375)

Thus the explanation of the absolute comic or the grotesque through a superiority over nature is doubtful because the grotesque harbors elements that bring humans closer to other animals. De Man seems to touch on this point—and even to provide the better explanation evoked by Baudelaire—when he states: "In a false feeling of pride the self has substituted, in its relationship to nature, an intersubjective feeling (of superiority) for the knowledge of difference" (p. 196). What would resist the fall and its effects would then be difference that would not entail invidiousness between humans and other animals. (One may here recall Baudelaire's observation in *Mon coeur mis à nu* that true civilization is to be found in the diminution of the traces of original sin.) But it is noteworthy that this difference without invidiousness is not for Baudelaire a pure disjunction. Nor is the distinction between the absolute and the significative comic purely disjunctive.

The latter point becomes evident when Baudelaire tells why Hoffmann is exemplary for him, for Hoffmann is not the exponent of the pure absolute comic but a hybridized type in whose texts the two forms of the comic interact:

> [Hoffmann] unites to the significative mockery of France the mad, sparkling, and light gaiety of the countries of the sun at the same time as the profound Germanic comic. . . . What quite particularly distinguishes Hoffmann is his involuntary—and sometimes very voluntary—mixture [*mélange*] of a certain dose of the significative comic with the most absolute comic. His supernatural and most fugitive comic conceptions, which often resemble visions of drunkenness, have a very visible moral sense: one might believe one is dealing with a physiologist or a very profound alienist [*médecin des fous*] who amused himself by clothing this profound science in poetic forms, like a learned man who would speak in allegories and parables. (P. 377)

Here one has a linkage between allegory, the absolute comic, and the significative comic having moral and social effects. Baudelaire also makes a point of noting the setting of Hoffmann's *Princess Brambilla:* "The joyous, noisy, and frightful Italy abounds in the innocent comic. It is in Italy, in the heart of southern carnival, in the midst of the turbulent Corso, that Theodor Hoffmann discerningly placed his eccentric drama, *The Princess Brambilla*" (p. 376). One might suggest that to the extent that the figures in the text, like Callot's "carnivalesque figures" (p. 375), float against a background that is precisely not the world, the text is both symptom and critique of the fate of the carnivalesque once it is separated from a more or less viable interaction with social institutions. In any event, radical defamiliarization should not be equated with pure transcendence.

I would conclude by stressing that Baudelaire sees the absolute and the significative or social comic as displaying the same difference as the school of art for art's sake and that of "interested" (or socially committed) writers. The terms of the argument indicate that the absolute comic is a higher type, but the difference or disjunction between the two is not pure or total. In Baudelaire's terms, it is all a matter of dosage or hybridization. More precisely, the "doubleness" of the human being is related to an undecidable status between two infinites, which are indicated by laughter itself: laughter is "simultaneously the sign of an infinite grandeur and an infinite misery, infinite misery relative to the absolute Being of which he [or it—the referent is unclear: it could be "laughter"] possesses the conception, infinite grandeur relative to animals. It is from the perpetual clash of these two infinites that laughter erupts [or disengages itself—*se dégage*]" (p. 373). Here Baudelaire powerfully conjoins a Pascalian image of the two infinites with a more modulated notion of necessary hybridization or variable dosage without postulating that the former points toward an impossible "authenticity" (indeed it generates what may be a mystified human sense of superiority) or that the latter designates a "degraded" compromise or relapse. And while the absolute comic—in its drunken hilarity, its *vertige de l'hyperbole,* its proximity to the unconscious, and its "lucid madness"—cannot be reduced to social and political considerations, neither can it be divorced from them. In any event, it is neither a one-sided, deadly ideal nor a vehicle for an unworldly pure fiction. From this complex perspective, one can sense the attraction of a self-consciously impossible desire for transcendence and purity, even—perhaps, in a certain modern context, especially—when it comes in what might seem to be the terminally secular form of technical linguistic analysis that "rigorously" marks and remarks the ultimate "unreliability" of all language, the impasses of "worldly" compromises, the fascination of the void, and the tempting "seductions" of reconciliation and symbolic meaning. Yet one problem Baudelaire would seem to leave us with is how to develop further his own conception of the intricate relation between "literary" or "artistic" texts and their pertinent but variable contexts, such as the carnivalesque. I have tried to suggest that a concern for the text/context relation need not be construed as an alternative to rhetorical inquiry or as an indubitable sign of a reductive if not foundational mode of thought. Indeed a certain kind of attentiveness to this relation may enable one better to trace—and to confront one's own implication in—the variable figurations of temporality (or historicity) in terms of a recurrent but changing formation combining symptomatic, critical, and at times transformative possibilities.[8]

In this essay, I have compared and contrasted two major contemporary critics on a crucial set of issues in interpretation related to the problem of temporality. M. H. Abrams receives briefer treatment than Paul de Man only because the main lines of his argument are clearly delineated and the details are too multifarious to be addressed in an essay. Although I have often departed from de Man's own protocols of interpretation, I have devoted to his densely intricate and influential essay the type of close, critical reading that he demanded in addressing texts and that is needed with respect to his own writings, as has become increasingly recognized.[9] Indeed the discrepancy in the space I have devoted to Abrams's large book and to de Man's relatively succinct essay is not without its irony (comparable, perhaps, to the irony in de Man's treatment of Baudelaire's *De l'essence du rire* and Wordsworth's brief poem). The crucial point on which I have insisted is that temporality is best seen as an intricate process of repetition with change—at times traumatically disruptive change that may nonetheless involve the return of the repressed. This process is oversimplified when it is resolved into an option between continuity and discontinuity, "symbolic" unity and "allegorical" disjunction. Furthermore, I have assumed that the interpreter is implicated in the temporal processes interpreted. In this sense, one cannot dissociate a substantive conception of time from methodological and theoretical issues or from one's very practice in writing. Indeed it is by attempting to think and work through uneven developments or intricate displacements and by relating them as cogently as possible to specific contexts of writing, reception, and critical reading that we can best address the problems in interpretation left to us by critics of the stature of Abrams and de Man.

PART IV

The Temporal Order of Social Life

The Constitution of Human Life in Time

THOMAS LUCKMANN

TIME IS CONSTITUTIVE OF human life in society. Of course it is also constitutive of human life in nature: *all* life is in time. But as a dimension of human life time is not only the matrix of growth and decline between birth and death. It is also the condition of human sociality that is achieved again and again in the continuously incarnated contemporaneity of face-to-face interaction. As an essential dimension of individual, interactional, and institutional existence, time confers an elementary historical character upon all human life in society. The emergence and subsequent interpenetration of several dimensions in the experience of time were a necessary condition for both the evolution and the ontogenesis of personal identity as the peculiarly human form of life.[1] As I will try to show, these dimensions are the *non*-identical diachronies of the body, of the ongoing synchronizations of self and other in face-to-face interaction and of "history." Personal identity arises at the intersection of these dimensions in individual consciousness.

"Time" has a peculiar status in sociological theory. In all the sciences theory is the product of (rather peculiar) specialized and systematizing communicative activities. But only the social sciences have a subject matter that is itself constituted in communicative activities: human life in society. In other words, the social sciences are necessarily reflective: second-order (i.e., theoretical) concepts refer to first-order (i.e., ordinary language, common-sense) concepts that guide everday behavior. Time is therefore evidently an important aspect of the reflexivity of the social sciences.

Sociologists have often complained that their discipline lacks a theoretical focus on time.[2] It must be admitted that the complaint is not wholly unjustified. It is true that basic efforts to comprehend the social nature of time were not entirely lacking in the sociological tradition. One need only think of Durkheim and his school and especially of Maurice Halbwachs's essays on collective memory and the social categories of time.[3] Equally important in a methodological context, although less known among sociologists, are Georg Simmel's investigations of the problem of historical time.[4] Sociality and time are the two most important elements in George Herbert Mead's theory of action,[5] and time is a pervasive concern of Alfred Schutz's protosociology, especially in its account of the constitution of meaning in experience and action, which in this regard is influenced as much by Bergson as by Husserl.[6] It is also true that interest in the temporal dimension of social processes was a matter of course in interpretive historical sociology, thus in the writings of Max Weber and also in those of Norbert Elias.[7] Furthermore, investigations of the temporal structures of institutions and organizations have begun to accumulate in recent decades. They are linked to the study of "timetables" and "scheduling" and are paralleled by a growing body of studies on various aspects of "careers" and "trajectories"—occupational, deviant, criminal, medical, marital—as well as of the social contexts of entire life-courses.[8]

However, it is the case that there is no common, clearly articulated perspective on time in sociology or, for that matter, anywhere in the social sciences.[9] "Time" is everywhere in human life and in the body politic. But the practical problems of the management of time in everyday life do not translate into a general theoretical effort in the social sciences. Time is so pervasive in the everyday life of the social scientist, so omnipresent as a condition of his theoretical and research activities and so basic a dimension of the object of his studies, that it is taken for granted—even as specific temporal aspects of particular institutions and organizations are closely inspected. In view of this, it is not difficult to understand the unease caused by the absence of an adequately elaborated common perspective on time.

Here I shall be concerned with the elementary issues of the relationship between time and sociality. But I shall also try to show the systematic connection between these issues and the specific problems of the temporal structure of organizations and the varieties of the social construction of temporal categories in language, calendars, and so forth.

The Constitution of Human Life in Time

The Body and Inner Time

In everyday life we take it for granted that one experience follows another. We also take it for granted that we can remember past experiences, some accurately, some vaguely. We even believe that we can "share" our past experiences with other people, recount them to others. We know that under normal circumstances we can do so with a degree of accuracy that suffices for most practical purposes. Correspondingly, we assume as a matter of course that we know what it is all about when others tell us of their experiences—even if we sometimes notice occasional involuntary or deceitful distortions of the past. In the "living" present, the present in which we are immersed with our bodies, we can grasp something that presents itself to us as past, as a recollection, just as well as something that presents itself as an actual ongoing perception or daydream. In fact, in some currently occurring experiences we even anticipate that we shall have occasion to recollect them later on. We note that they are memorable. We often look at an ongoing experience from the currently anticipated future, before that experience recedes into the past.

Moreover, in everyday life there is one kind of experience that has an even more complex temporal structure: an experience that is rehearsed in the ongoing present before it has "really" happened—and then, other things being equal, willed to happen. With Schutz[10] we may call such experiences "actions": sequences of experience that were projected earlier and whose meaning is constituted in (partial) fulfillment or (partial) frustration of the original project. However, actions, although they can be performed by a solitary individual, are intersubjective in origin and will be considered later; we shall first look at experiences with a simpler temporal structure.

Our body roots us in the present by its actual "inner" and "outer" perceptions. But the present in which we live, and of which we are concretely aware, moves on. We say that time "flows." It is to this inexorable flow in which one experience succeeds another that we attach the notion of time in the "natural" attitude of everyday life. Yet, even in a prephilosophical, common-sensical attitude, we also seem to be able to step outside this flow. This is a powerful, operative illusion deserving of closer scrutiny.

In his investigations of inner time, Husserl has shown that the unity of the stream of consciousness over time is constituted in the continuous syntheses of what is given just now with what was given as just now a moment ago and with what is not yet a just now, although about to become now momentarily.[11] These fusions of the actual, impressive phase of consciousness with retentions of a previous impressive phase and protentions of a coming impressive phase occur in continuous *automatic* syntheses, not in intentional *acts*. Inner time is the elementary form of passive intentional processes: an irreversible sequence. Every ongoing experience is located in what William James most appropriately called a "specious present," and it has two temporal horizons: one is a backward horizon behind which everything that is now present is constantly receding. I recognize that which has receded as my past. There is also a forward horizon from which consecutive experiences move forward into the ongoing present. It may or may not be what was anticipated a moment ago in what was the then current impressive phase. But whatever it turns out to be, it also recedes into the past.

This is the *elementary* temporal structure that characterizes all intentional processes of human consciousness. But, although everything else rests on this temporal basis, there are other levels of time in human experience. Husserl pointed out that one normally "lives" in one's intentional activities, that one is occupied with the intentional *objects* of these activities, not with the activities themselves.[12] In order to grasp either these activities or what was presented in them but is no longer presented "in person" in the present experience, it is necessary to turn to them reflectively—and that means, retrospectively. The acts of reflection and the retrospective glances are inevitably embedded in the ongoing present. But the intentional objects, although apprehended in the present impressive phase of inner time, do not present themselves as actual and present. They present themselves as past, and this temporal subscript ("past experience") is an intrinsic part of their meaning.

All experiences are, of course, constituted in time, step by step: polythetically. If I return to them retrospectively, I may follow the path of their original constitution and reconstruct them as they were constructed, once again step by step. I *may* follow this path, but I *need* not do so in all cases. The meaning of many experiences may be grasped in one glance, monothetically. The meaning of *some* experiences is, however, intrinsically polythetic and must therefore be reconstructed polythetically. Musical themes must be replayed, historical events must be narrated, if one wishes to grasp their nature.

Inner time—the point has been forcefully made by Bergson and William James—cannot be adequately segmented into equal units without doing violence to its intrinsic articulation. Inner time is the form of consciousness as continuous experience; it is not a measurable object in space. If inner time has units, they cannot be stretches of the *res extensa;* they must be sequences of the *res cogitans*. It is obvious that the rhythm of these sequences is naturally rooted in the human body with its alternation of wakefulness and sleep. William James spoke persuasively of "flying stretches" and "resting places."[13] However, although we are automatically aware of the rhythmic structure of inner time, we do not normally think about it. If we do think about it at all, we cannot help but use the categories of time and space that are socially objectivated in various ways. Such categories are internalized by the members of a society and are thus superimposed more or less "artificially" on the elementary rhythms of inner time. These categories are of course not originally those of inner time itself. They originate in social interaction, and they are determined by the requirements of social organization (communication, work, and political institutions). They are socially preserved in objective (semantic-syntactic) forms, and they are socially transmitted.

It is a general trait of all conscious life that in each successive actual phase the immediately preceding phase is automatically retained, just as the next impressive phase is automatically anticipated. Retention, impression, and protention are fused in a flow of ongoing impressive phases, without much intrinsic segmentation of experience. They continue to flow in a characteristic body-bound rhythm. Upon this flow are superimposed both subjectively motivated shifts of attention that may slow it down or speed it up and externally imposed changes in experience. Attentiveness may change when an individual "shifts gears" from mere experience to action, when he proceeds from unimportant to urgent action, and, of course, when he moves (or is moved) from one domain of reality to another (e.g., from waking to dreaming).

The rhythms of inner time are the basis of experience, and all other structures of time in human life are erected upon it. The latter, however, do not *originate* in the (pre-predicative) inner time of a solitary self. They originate in social interaction. Reflective turns to prior experience and projects of future action that normally depend on highly socialized categories are, of course, *located* for each individual in his inner time. Each successive present experience, that ongoing synthesis of impression, retention, and protention, is surrounded by a horizon of definitely past experience, necessarily closed, and of potential future experience, necessarily open-ended but nonetheless anticipated as going to have not only a typical beginning but also a typical end. Each successive experience is thus, in a manner of speaking, the (partial) fulfillment or the (partial or full) disappointment of what was anticipated earlier and—in the case of action—of what was projected earlier.

Social Interaction and Intersubjective Time

The rhythms of inner time are the immediate "link" between body and solitary consciousness. While the individual remains solitary and as long as he is neither working nor thinking to some purpose, his awareness is regulated by these rhythms. But daily life does not offer much solitude. Periods in which the individual is not attending to practical affairs in some way—working, talking, thinking—are rare. The world of everyday life is a social world of crisscrossing actions. The temporality that characterizes the world of everyday life is therefore not the same as that which characterizes the daydreams of the solitary individual.

The inner time of those who live and act together must be reciprocally adjusted. Such adjustments of course need be neither continuous nor permanent, but they must be undertaken at the very least on all occasions when the individuals engage in social interaction. This evidently covers important stretches of daily life.

The adjustments that are required in the coordination of social interaction are of two kinds. One kind—which we may consider to be elementary—consists in the synchronization of two streams of consciousness. The other—which we may consider to be derivative—results from the superimposition of socially objectivated temporal categories upon concrete social interaction. Empirically the two kinds appear in various combinations.

Synchronization of two streams of consciousness may be said to occur when the body-bound rhythms of two individuals are placed in a "parallel" course so that both are aware of sharing the same experience (e.g., of a bird flying by). The synchronization of two streams of consciousness can only come about when the individuals are situated in bodily presence of one another. (I shall disregard the complexities of partial and quasi-synchronizations achieved by various technological means.) In face-to-face situations the body of one individual is experienced directly by the other, and vice versa. To partners in a face-to-face situation the bodies of fellow human beings are a rich field of expression, expressing an individual's actual, ongoing consciousness: his momentary perceptions, intentions, fears, wishes, and so forth. And as the other perceives and interprets the first individual's expression, he experiences his own perceptions and interpretations as simultaneous with the experiences expressed by the first individual. In a face-to-face situation, expression and perception of expression are co-related and reciprocal. The individuals, of course, do not have the same experiences, but they certainly do share experiences.[14]

Once synchronization of two streams of consciousness is achieved, actions originating with different individuals can be geared into one another so as to form a unitary course of social interaction. Successful interpretation of cues is a necessary condition for the accomplishment of social interaction. Under normal circumstances such interpretations are achieved "automatically" as, for example, in reciprocal gazes.[15] That is not to say, however, that even fairly simple coordinations of time above a most elementary level do not represent an early problem of "socialization" and a need to be taught and learned.[16]

Synchronization of two streams of consciousness is only required in face-to-face, direct social interaction. Even there the degree of synchronization that must be achieved varies. Building a house together differs in this respect from making love, fencing differs from a liturgical responsory. Different degrees of exactness are required for the timing of different kinds of social interaction. In some kinds of work, as, for example, in ritual sequencing of procedures or intimate bodily interaction, the temporal aspects of experience and action remain in focus; synchronization here may always become problematic. In highly routinized social interaction face-to-face synchronization is still necessary, yet may be achieved in a correspondingly routinized fashion.

In the temporal coordination of more or less routinized social interaction the required adjustments are usually accomplished with the aid of ready-made categories. In face-to-face situations these categories are of course superimposed upon an elementary synchronization of streams of consciousness. More precisely, they are superimposed on a mediating layer of interactional procedures, such as the taking of turns, that themselves rest upon the synchronized rhythms of inner time. The categories themselves are anonymous and abstract: they structure time in an obligatory fashion for "everybody." "Ready-made" means that the categories have a degree of social objectivity and that they have been socially transmitted to individuals rather than having been created by them in the process of social interaction. And "structuring time" means that stretches of typical kinds of action and experience are given objective status by being differentiated from (and compared to) one another as well as from events that are independent not only of the rhythms of inner time but also of social interaction, that is to say from external natural events that, of course, are thus endowed with considerable social significance.

It should be noted again that the time that governs daily life is intersubjective: the concretely experienced, "lived" time of direct social interaction. It is not the time of abstract social categories. On the other hand, it is not the time of "pure" intersubjectivity, that is, of original, prepredicative, precategorical synchronizations of two streams of consciousness. The time of everyday life is *socialized* intersubjective time.

Temporal categories are of course ready-made only from the point of view of any given individual born and socialized into a particular society in a particular epoch. But, evidently, the categories have not made themselves. They were made, and are constantly being remade in the very processes of social interaction whose temporality they help to regulate as soon as they are available. One may take it for granted that synchronization of social interaction could and would come about, no matter how laboriously, even in the interaction of individuals who were *not* socialized into the same world view and thus could not employ the same categories. The "objective" categories of time, on the other hand, cannot be imagined to have come into existence without some prior social interaction. Some synchronization of the inner time of the actors must be presupposed for the accomplishment of any social interaction. The objective categories are *socially* objective: social categories of time thus point back to original intersubjective coordinations of interaction sequences accomplished in precategorical synchronization of two streams of consciousness.

The adjustments of inner time necessary for the successful accomplishment of various kinds of social interaction are typified by the actors on some level of awareness. The typifications are rudimentary cognitive schemes (of the "If . . . then . . . " kind) and therefore necessarily "abstract" to some degree. They are applied to future interactions of the same kind. In the process of application to concrete interaction they reacquire "life." If the typifications turn out to be problematic in some way (e.g., in being "overgeneralized" to the wrong kind of interaction, individual, etc.), they become a topic for reflection and some systematization on the part of the actors. If experts are available in a given social structure, further systematization tends to be turned over to them. The solutions to the problems of temporal coordination are socially objectivated: categories of time and temporal categories of interaction and experience are articulated in words, ordered numerically, represented symbolically, and so forth. They become "abstract": anonymous and independent of the inner time of any living individual.

Once categories of time become objectivated as parts of a social stock of knowledge, they are routinely transmitted to the members of a society. Consequently, social interaction can be coordinated without original acts of synchronization and without original reciprocal typifications of action sequences. Complex social interactions can be organized on the level of institutions with the use of abstract social categories of time in total disregard of the rhythms of inner time that instill life into the synchronizations of two streams of consciousness in direct social interaction.

It is obvious, nonetheless, that the temporality of *daily* life arises in the coordination of intersubjective sequences of interaction, and that this coordination rests upon the synchronization of the inner times of two or more individuals in bodily presence of one another. It is also obvious that some kinds of interaction continue to require elementary synchronizations of inner time and some elementary reciprocal attentiveness of the actors even if social categories of time are routinely superimposed on them. In immediate face-to-face social interactions the social categories of time provide external temporal settings. But interaction still requires constant "tuning-in." The temporality of daily life, as effectively as it may be structured by abstract, socially objectivated categories, is the intersubjective temporality of immediate social interaction and rests on the synchronization of the rhythms of inner time among men and women.

Time and the Social Stock of Knowledge

Inner time thus has its "location" in the body of the individual human organism, intersubjective time has its location in direct, face-to-face social interaction. The social categories of time, on the other hand, have no comparably concrete location; they are not embodied. Nonetheless they can be said to be located in the social stock of knowledge of a given society at a given time.

A social stock of knowledge has a virtual existence in the minds of the members of a society who share that stock. But the social stock of knowledge does not merely have such virtual existence. It exists also in continuous or recurrent actualizations: in the experiences and actions for which the elements of knowledge are relevant.

A special and most important instance of actualization of the elements of knowledge (in the present case: the social categories of time) is to be found in the social transmission of such elements in practical exemplification and instruction. In addition to such primary—verbal, gestural, mimetic, and so forth—actualizations of elements of the social stock of knowledge in initial transmission and subsequent everyday use, a secondary objectivation may be found in some societies in the form of notations, scripts, and instruments. The obvious examples of such objectivations in the case of the social categories of time are of course calendars and clocks.[17] Yet not only the primary actualizations of the social categories of time are acts of consciousness or include acts of consciousness; the secondary actualizations, too, are embedded in acts of consciousness. The use of calendars and clocks presupposes the elements of knowledge that they embody.

The sequencing of the daily routines of social interaction requires categories that are explicitly temporal (such as calendars) or define beginnings and ends in indirectly temporal ways. The categories mediate between the small-scale interactional temporal structures that, at least in principle, could be negotiated intersubjectively (although they are normally not so negotiated in fact), and large-scale institutional structures of time that are socially objectivated on a high level of anonymity and whose jurisdiction may transcend the life span of an individual. The social categories of time serve to stabilize the recognition of typical beginnings, durations, and ends of typical experiences and interactions. They are historically variable, although, presumably, limits are set to the range of variation by the elementary structure of human consciousness on the one hand and the basic requirements of social organization on the other.

In comparison with the body-bound rhythms of inner time the socially objectivated temporal categories seem abstract and remote from experience. Nonetheless, they are concretely interwoven in the ordinary business of everyday life. The time of daily life is socialized by categories derived from the social stock of knowledge, yet it retains a concrete basis in experience, a basis that consists of ongoing intersubjective synchronizations and the systoles and diastoles of inner time. Furthermore, the dividing line between the genuinely abstract categories measuring duration independently of interaction ("clock time") and the far more concrete categories that shape the flow of social interaction by defining openings, closures, interlockings, and overlaps is not sharply drawn in preliterate societies. In such societies socially objectivated categories of time are therefore very visibly involved in the "lived" time of daily life. Temporal categories are also embedded in the myths and symbols that in such societies link the routines of ordinary experience with extraordinary realities and their "times."

Abstract categories of time became fully divorced from immediate experience and gained the status of quasi-ideal systems only in a few complex literate societies. Along with spatial categories they were first systematized in the early civilizations by experts of various kinds and found an institutional location as special knowledge. But although the "theories" of time and space thus developed no longer belonged to knowledge that was generally distributed in society, the *practical* use of the measures of time (and space) of calendars, clocks, watches (and yardsticks, blueprints, maps) required little or no specialization. It became part of daily routine. Thus even highly abstract categories of time—wherever they had developed—penetrated the interactional structure of daily life.

In these societies abstract temporal categories serve to regulate not only large-scale and anonymous organizations but also many temporal aspects of immediate social interaction. They do so just as effectively as temporal categories "closer" to intersubjective experience regulate interaction in face-to-face communities. The development of abstract categories of time stood in systematic relation to the requirements of the large-scale political organization of action, especially of work, and war in early civilizations.

Biography and Historical Time

In addition to the categories that structure the temporal dimensions of social interaction and the increasingly more abstract measurements of time in complex organizations, social stocks of knowledge contain another important set of temporal categories, categories by which an individual course of life is fitted, from birth to death, into a time that transcends it.

It is not surprising that categories in this set (one might call them biographical schemes) bear a closer resemblance to the diachronic interactional categories that differentiate aims and projects, define beginnings and ends, and coordinate interlocking phases of action than to the synchronic categories that measure duration (and provide a spatialized taxonomy for it). Both interactional categories and biographical schemes serve to order sequences of action in time. But there is a substantial difference in the temporal span on which the ordering occurs. Interactional categories integrate sequences of action within the individually recurrent and short stretches of daily life; biographical schemes integrate sequences of (temporally already preordered) action within an overarching, individually *non*recurrent "long" sequence, the entire course of an individual's life.

There is another, subtler difference. Interactional time categories are, in a sense, diachronic—but their function is, first and foremost, to *synchronize* interaction. The sequential ordering is primarily intersubjective and *social,* although this also involves an ordering of individual routines in daily life. Biographical schemes, on the other hand, are diachronic in the full sense of the word, and their basic function is to integrate short-term into long-term temporal sequences—and the long-term sequences are *individual* rather than social. They are socially constructed as all other elements of a social stock of knowledge—but they are tailored to fit an individual's lifetime rather than institutional times.

Biographical schemes endow the meaning of short-term action with long-term significance. One might say that this is a matter of "reverse telescoping": what looms large when seen with the naked eye recedes into the background if one looks at it through the wrong end of the binoculars. The meaning of daily routines does not disappear in such telescoping but is set in relation to the vaster background of an entire lifetime. This, however, is not the only function of biographical schemes. In view of the intersubjectively gained knowledge of the finitude of human life, they also endow some actions and experiences with special significance and make them stand out sharply from daily routines. One might describe the procedure involved here as proper "telescoping." Something that perhaps would not appear to be of special importance by itself is brought forward by the magnifying lens of the telescope. The lens is ground and polished by a hard diamond: human, intersubjectively gained knowledge of the finitude of human existence.

Biographical schemes cannot—entirely—avoid the problem of death. They attempt to "solve" it by linking individual existence to something that transcends it. From the point of view of the individual, time that transcends the human lifetime may be said to be "historical." There is no denying that simple as well as complex theories about his location in "historical" time profoundly influence an individual's conduct of life. I shall return to this point again. Historical time in the sense proposed here is as universal in human societies as interactional time and inner time.

Biographical schemes thus perform two kinds of closely related functions. They "convert" short-term into long-term relevance and vice versa. The meaning of a large variety of ordinary, everyday actions and experiences is subordinated to biographical significance, and biographical significance is infused into certain kinds of actions and experiences directly. Both procedures, hierarchization and dramatization, fuse short- and long-term sequences of action into meaningful components of an entire course of life. They do so by locating an individual course of life in relation to something that transcends the individual's lifetime. That "something" may be primarily the time of a transcendent social entity: family, kin, or a larger social whole such as nation, class, and the like. It may be transcendent nature or it may be an *individual* transcendent "timeless" time: eternal life.

Biographical schemes link large stretches of a typical individual's life and even his entire life to transcendent, historical times. As elements of a social stock of knowledge transmitted in socialization, biographical schemes trace individual paths through the social and historical world. They consist of formulaic versions of obligatory or possible lives or parts of life, with some instruction as to how the parts are to be put together in order to form whole lives. It may be said of interactional categories that they already tend to combine so as to produce formula-like wholes with normative, explanatory, and legitimating connotations. Evidently, this is so in increased measure in the case of biographical schemes. Biographical schemes are explanatory, legitimating, and normative "models."

It hardly needs to be stressed that both the "contents" of the biographical schemes and the communicative forms into which their "contents" are poured—most notably the oral and literary, everyday and religious and poetic genres—vary from society to society and from epoch to epoch. The range of variation is wide. Biographical significance inhabits such diverse notions as honor, dignity, sanctity, fairness, success, and self-fulfillment, and links them to essentially temporal categories such as natural, social, and individual seasons, feasts and liturgical cycles, and superordinated categories such as Fate, Providence, Salvation, Chaos, Cosmic Order, and Nothingness. Biographical schemes may be elaborated in independent genres such as legends, *vitae sanctorum,* etiological myths, confessions, diaries, and autobiographies, or worked into catechetic and other pedagogical subgenres (e.g., sermons and schoolbooks).

Ordinarily the word "biography" refers to a fully developed, historically and socially restricted literary genre in which the course of an individual life is narrated, written down, and published in some form. For my part, I am using the term in an extended sense to refer to any socially objectivated oral or literary scheme or model for the course of an individual's life. I think such schemes necessarily contain a narrative core, that is, a sequential arrangement of events and actions, and formulaic condensations, that is, typical ordinary and reverse telescopings.[18] How such models become elements of the social stock of knowledge, how and to whom they are diffused, and in what kinds of communicative processes, whether they form independent genres or are fused into essentially nonbiographic genres, whether they enter socialization in pedagogical blocks or obtain a certain autonomy, all these are matters that vary from society to society and from epoch to epoch.[19]

Biographical schemes form the basis for individual projects of life, for the planning, evaluation, and interpretation of daily routines as well as of dramatic decisions and critical thresholds; they provide a scaffolding for partial and full reconstructions of a life's course for legitimatory, instructive, or other purposes. The categories and interpretive schemes by which human beings order their course of life prospectively and retrospectively—and in terms of which they reconstruct, narrate, and legitimate their lives to others and to themselves—are evidently not categories of inner time. They are categories that are somewhat removed even from the structures of intersubjective, at least partly negotiable interactional time. In innumerable anonymizing and idealizing communicative acts they have attained the status of cultural objectivity and tend toward systematization.

Nonetheless, the temporality of daily life and the temporality of an individual course of life embedded in historical time are linked in various ways. In reflection and retrospection, in planning and projecting, in reconstructing and legitimating, biographical schemes are superimposed on the routines of daily life. The polythetically constituted sequence of many a day's experiences and actions can be thus retrospectively summed up in narrative formulae (of high evaluative content) in a manner that comes close to a monothetic grasp of them. On the other hand, both projects for the distant future and interpretations of one's past are necessarily inserted into the small-scale temporal articulations of daily life. Biographic schemes are particularly important as constituents of the *conscious* and explicit dimensions of personal identity. Its categories claim a high degree of objectivity—"this is the way it is"—and they retain a subjective reference point: they objectivate what to the individual is *his,* his unique sequence of experiences and actions, sedimented in *his,* his unique memory. In addition to those elements of an individual life that are predetermined by a social structure and those that are simply contingent, there are those that are the result of his own actions—and *those* are guided by biographical schemes that were internalized by him.

Historicity, Historical Consciousness, and a Modern Fissure of Time?

"Historicity" is a universal characteristic of human life in society. When individual courses of life become "biographies," when they gain contours by being situated in the transcendent time of institutions, by being inserted into ancestral lines and oriented with respect to (dynastic) successions in the body politic, individual actions no longer automatically serve to maintain or change a given social order. They *may* maintain it—as they may transform or even destroy it. Concrete situations are no longer determinants of human action (as they, in combination with genetic programs derived from a biogram of the species, tend to be in other species). The meaning of individual action is not encompassed by an individual's life span.

This much can be said about the anthropology of time without becoming enmeshed in the Heideggerian net of an (*Eigentlichkeits-*) ontology of being-toward-death. But even if historicity is a universal trait of human life in society, what about historical consciousness (which, perhaps, is more closely connected to an awareness of mortality)?

Great historians, among them Jacob Burckhardt and Ranke, philosophers such as Jaspers, and even scholars familiar with preliterate cultures, for example, Mircea Eliade, have thought of archaic society as unchanging or, if they were aware of the fact that this is not so, insisted that only explicit and nonmythical knowledge of "origins," of change, and so forth makes for "true" historical consciousness. In fact, the ruling opinion was that historical consciousness is possible only in literate societies.

It is easy to show that this position is untenable and that it can be reduced to the absurd and probably unintended argument that history is an artifact of the historiographers. However, even if historical consciousness of some sort as well as the historicity of human existence can be assumed to be universal, it does not follow that different kinds of historical consciousness could not have developed. Everyone is conscious of the limitations of human experience and of human life and aware of the fact that life and experiences are embedded in something that transcends them in time and space. Furthermore, human beings everywhere acquire knowledge about their own finitude as they observe the mortality of their fellows. Nonetheless, such knowledge does not automatically scale down the potential of expectations to the measure of individual finitude. This is a problem for all societies, to which they have responded in different ways. Socially constructed models of restraint and resignation, as well as of small and great expectations and hopes transferred to other levels of reality, of individual and collective utopias and eschatologies, have dealt more or less successfully with this specifically and intrinsically human "fissure" of time.[20]

Many different circumstances influence the social determinants of "historical consciousness." The structure of the social stock of knowledge (from simple to complex) and the degree of institutionalization characterizing the social transmission of knowledge (from age-grades to Babylonian scribal schools to modern education systems) constitute one important group of factors. Another consists of the repertoires of "historical" versus purely "fictive" genres of narration, commemoration, and so on that are available in a society, and the degree of differentiation of such genres. No doubt literacy is a particularly significant aspect of the "technology" of preserving and transmitting traditions, and it also changes the character of canonization and censorship, which, understood in a broad sense, determine the selectivity of tradition and thus of historical consciousness in preliterate as well as literate societies.[21]

It hardly needs to be stressed that all these factors and combinations of factors are at least partly determined by the general type of social structure in which they are operative. Hunters and gatherers (for example, the Pygmies of the Ituri forest) have a less clearly articulated sense of tradition than even the acephalous societies segmented along clan principles (as in some African tribes) for which descent and ancestral continuity may play an important role. Historical consciousness is more clearly articulated in societies with a central political institution (e.g., chieftainship). A particularly important, possibly qualitative change of the nature of historical consciousness occurs in the conjoint institutionalization of power and religion, as in the divine kingships of the Ancient Middle East and especially in Pharaonic Egypt.[22] A shift of perhaps comparable magnitude occurs with the historicization of the future for which the basis was established in Ancient (especially exilic) Judaism and which characterizes the Christian eschatology and its secularized descendents in the philosophies of the Enlightenment, Hegelianism, and Marxism.[23]

But this is a well-known story. I merely wished to reinforce the point that one variant of historical consciousness must not be taken as the only kind there is—nor be made the measure for the others even if their existence is halfheartedly admitted. Furthermore I think that it is precisely the clear understanding of the universal substrate of "historicity" in human action and of "historical consciousness" in the pretheoretical orientation in the world that helps us to discern and to interpret its fascinating variations and articulations.

Fascinating and, on occasion, perhaps fateful. There may be a line in our own tradition that leads from the *ars longa, vita brevis* of the ancients to the resignative sentence in Montaigne's *Essays,* "Le temps me laisse; sans lui rien ne se possède,"[24] to the modern invention of "midlife crises." Hans Blumenberg devoted his latest book to a magisterial study of the continuities and changes in the Western experience of time. His findings cannot be easily summarized; the story is too complicated to allow for formulaic simplifications. However, one may represent his main thesis in the metaphor he himself uses: the gap between the two blades of the temporal scissors is widening—where one blade is the time of individual existence in the world and the other blade is world time.[25] It is unlikely that the scissors were ever completely closed. In this world human life is an intersection of diachronies—in fact, the identity of experience and action in these diachronies constitutes personal identity. But Blumenberg makes a very strong case for his thesis that in our present situation the time-scissors have opened wider than ever before or anywhere else.

Synchronizing Individual Time, Family Time, and Historical Time

TAMARA K. HAREVEN

HISTORIANS HAVE BEEN intensely aware of the importance of time as a major frame for social change, but they have discovered only recently the importance of time as a *phenomenon* or product of social change. The concept of time as a historical construct has been discussed almost exclusively by historians concerned with the transition from task- and season-related work patterns and ways of life to industrial time discipline. For example, the importance of modern industrial time schedules has been a recurring theme in labor history, where it was first presented by E. P. Thompson. Thompson emphasized the introduction of the industrial time clock and discipline as a dramatic and traumatic watershed in the history of Western society.[1] Until recently, less attention has been given to issues of time and timing as manifested over the individual life course in relation to changing historical time.

Over the past decade and a half, however, research in the history of the family and the life course has introduced an additional dimension into the understanding of time in relation to historical change: The historical study of the family has demonstrated that in its interactions with historical change, the family often followed its own time clocks. Even though the family responded to larger forces of historical change, it formed its responses and adjustments in relation to the time clock dictated by its own cultural traditions and needs, within the constraints of the larger social and economic structures. For that very reason, the periodization of family history does not fit the neat categories of established historical periodization (preindustrial, industrial, postindustrial; or premodern, early modern, modern).[2]

Hence, the recent findings in the history of the family challenge the basic assumptions of modernization theory that social change involves a linear process, whereby people shed their traditional values as they become "modern." Research findings in labor history, social history, and family history over the past decade have converged to show that even people caught up in rapid social change carried over many aspects of their traditions, and drew upon their traditional values and networks as resources in their process of adaptation to change. Thus, the timing of the adoption of "modern" patterns of behavior varied considerably from one group to the next, even within the same historical time period.

The historical study of the life course offers an opportunity to understand the issues of synchronization of individual time, family time, and historical time. It enables one to study these interactions on the behavioral level through the timing of life transitions, and on the perceptual level through individuals' own perceptions of their timing of transitions in relation to the social time clocks. By examining the various aspects of timing on the individual and familial levels in relation to changing historical conditions, the life-course approach enables us to understand the ways in which individual lives were synchronized with the larger processes of social change.[3]

This paper examines the impact of new concepts of time on the social clocks that individuals and families followed in the context of changing historical time. The type of "time" addressed here is not chronological in the strict sense. Its essence is *timing*—meaning coincidence, sequencing, coordination, and synchronization of various time clocks, individual, collective, and social structural. The first part of this paper defines the concept of "timing" from a life-course and historical perspective. The second part compares the patterns and perceptions of timing of three different cohorts in the United States. The third part compares these patterns with those in Japan.

The Definition of Timing

An understanding of social change hinges to a large extent on the interaction between individual time and social-structural (historical) time. In this interaction the family acts as an important mediator between individuals and the larger social processes. Under certain circumstances, individual timing harmonizes with family timing; under others it is out of synchrony with family time. The tension between individual timing and the collective timing of the family (especially in transitions such as leaving home, launching a career, or getting married) has also varied considerably over historical time.

The form and content of this interaction of different time clocks has changed over historical time and has varied in different cultural settings. Timing and timeliness are defined differently in various cultures and under different historical circumstances. Even within a specific time period a variety of definitions of timeliness may coexist—whether individuals or families are "early," "late," or "on time" in the scheduling of individual or collective family transitions may vary among different ethnic groups even within the same society during the same time period.[4] In order to grasp individuals' interpretations of their timing of such transitions, one needs to understand, therefore, the cultural and historical contexts within which specific life transitions are timed. The definition of the timeliness of certain life transitions depends on different societies' cultural constructions of the life course. Definitions of timeliness derive from individuals' inner expectations, and from those of their family members and their peer group. Timeliness is also shaped and defined by one's culture or subculture and by the values governing "timing" in different social and cultural contexts.[5]

For example, in Japan as in the United States, the "timeliness" of transitions is a major source of cultural concern. The meaning of timeliness in Japan is somewhat different, however, from its meaning in the United States. In Japan, the term *tekirei* designates set norms of timing—the ages "fit" for accomplishing various life transitions. Individuals thus define their timing of marriage, work careers, and other roles in relation to the set norms of timing defined in *tekirei*. *Tekirei* also means the orderly sequence of life transitions. In Japan more than in the United States, the orderly and fixed sequencing of life transitions, especially in the transition to adulthood, has been considered the proper normative order for completing school, starting one's first job, and marrying. This sequence is followed by more than 90 percent of each cohort, a fact indicating a uniform tendency in timing. By contrast, the normative order is followed by only 60 percent of each cohort in America.[6]

Until recently, the timing of life transitions in the United States was less related to age norms. Family need took priority over age norms in determining the timeliness of life transitions. Hence, transitions to adulthood (leaving home, starting to work, getting married, setting up a separate household) did not always follow an orderly sequence, and took a long time to accomplish, because they were timed in relation to familial needs, rather than in relation to age. Only since the post–World War II period has the timing of these transitions begun to adhere more strictly to age norms.[7]

The concept of timing over the life course involves the movement of individuals from one state to the next, rather than the segmentation of the life course into fixed stages. Such movements have been defined in life-course research as transitions. "Timing" thus designates when an event or a transition occurs in an individual's life in relation to external events, whether a transition conforms to or diverges from societal norms of timeliness, and how it relates to other people traveling with the individual through life or through certain parts of life. Thus the variables used in the examination of timing are relative rather than absolute chronological categories. They are perceptual as well as behavioral for those subject to them and for their associates.

Characteristics of Timing over the Life Course

Three features of timing are central to the understanding of changes over the life course: first, the timing of transitions over an individual life path, particularly the balancing of individuals' entry into different family and work roles and their exit from these roles; second, the synchronization of individual transitions with collective family ones; and third, the cumulative impact of early life transitions on subsequent ones. In all these areas the pace and definition of "timing" hinge upon the social and cultural context in which transitions occur.

(1) *Individual Timing.* In the individual life trajectory the crucial question is how people time and organize their entry into and exit from various roles over their life course: for example, how they sequence transitions in their work life and educational life with respect to transitions in their family life, how they synchronize and sequence transitions in the context of changing historical conditions in areas such as entry into the labor force, leaving home, getting married, setting up an independent household, and becoming parents.[8]

(2) *Synchronization of Individual with Collective Timing.* On the familial level, timing involves the synchronization of individual life transitions with collective family ones, and the juggling of family and work roles over the life course. The familial configurations in which individuals engage change over their life course. Along with these changes people time their transitions into and out of various roles differently. Age, although an important determinant of the timing of life transitions, is not the exclusive variable. Changes in family status and in accompanying roles are often as important as age, if not more so.[9] Thus, the synchronization of individual transitions with collective family ones is a crucial aspect of the life course,

especially where individual goals are in conflict with the needs of the family as a collective unit.[10]

(3) *Cumulative Impact of Life Transitions.* The third feature of timing is the cumulative impact of earlier events on subsequent ones over the entire life course. Early or late timing of certain transitions affects the pace of subsequent ones. Events experienced earlier in the life course can continue to influence an individual's or family's life path in different ways throughout their lives.[11]

Historical forces play a crucial role in this complex cumulative pattern. Historical conditions that individuals encounter in their earlier life history have a direct impact on their lives; and these events in turn may have an indirect impact on the later years of life. This means that the social experiences of each cohort are shaped not only by the discrete historical conditions prevailing at the time of those experiences, but also by historical processes that had shaped their earlier life transitions.[12]

(a) Generational transmission of timing: The impact of historical forces on the life course does not stop with one generation. Each generation encounters a set of historical circumstances that shape its subsequent life history and transmits to the next generation both the impact that historical events had had on its life course and the resulting patterns of timing. For example, among groups of parents and children that experienced the Great Depression, delays or irregularities in the parents' timing caused by that experience affected their children's timing. In this case, historical events had an impact on the children's lives on two levels: directly, through the children's encounter with these events, and indirectly, in the "ripple effect" of these events from the parents' generation to their children.

(b) Impact of long-term historical events: In addition to the immediate historical events that each cohort experiences, long-term historical change has a critical impact on timing over the life course in several areas. In the area of demographic behavior the timing of marriage, fertility, and mortality patterns shape changing age configurations within the family.[13] Similarly, external economic changes in the opportunity structure affect changes in the timing of entry into the labor force, and, ultimately, retirement. Institutional and legislative changes, such as compulsory school attendance, child labor laws, and mandatory retirement, shape the work-life transitions of different age groups, and eventually influence their family life as well.

Cultural norms governing the timeliness of life transitions (being "early," "late," or "on time") and norms governing familial obligations also shape individual and collective family timing. In all these areas, historical and cultural differences are critical. Particularly significant is the convergence of socioeconomic and cultural forces, which are characteristic of specific time periods and which influence directly the timing of life transitions and perceptions of the life course.

For example, "middle-age crisis" was a relatively recent invention by popular psychology in American society. It was attributed to middle-class women in particular in describing the problems connected to menopause and the "empty nest" in mid-adulthood. "Middle-age crises" were not widespread, however. Where they existed, they were caused, it has now been explained, by social stereotypes rather than biological or familial realities. Over the past decade and a half a considerable body of feminist psychological literature has placed "middle-age crisis" in its proper perspective by exposing the cultural and "scientific" stereotypes that created the concept.

Similarly, the definition of "adolescence" as a distinct stage of life was not a universal phenomenon. It emerged in a specific historical time period. The concept of adolescence was developed in the late nineteenth century in order to define the social and cultural characteristics accompanying stages of psychological development related to puberty. But even though all teenagers undergo a psychobiological transition in puberty, the cultural and social phenomena associated with it are not uniformly experienced in all societies; nor did they receive the same recognition in the past.[14]

In summary, any examination of transitions in the life of individuals is contingent on several factors: the place of such transitions in an individual's life in relation to other transitions; the relationship of an individual's transition to those experienced by other family members; and the historical conditions affecting such transitions. For this reason the differences among cohorts in experiencing and defining transitions is crucial for understanding the impact of social change on the life course.

Transitions and Turning Points

Transitions are processes of individual change within socially constructed timetables, which members of different cohorts undergo. Many of the transitions that individuals experience over their work and family lives are normative; others are critical or, at times, even traumatic. Transitions are considered "normative" if a major portion of a population experiences them, and if a society expects its members to undergo such transitions at certain points in their lives in conformity with established norms of timing. Under certain conditions even normative transitions might become critical ones, and might be perceived as turning points. *Turning points* are perceptual roadmarks along the life course. They represent the individuals' subjective assessments of continuities and discontinuities over their life course, especially the impact of special life events on their subsequent life course. In some cases, turning points are perceived as critical changes, in other cases as new beginnings.[15]

A central question in life course analysis is therefore: What transitions are considered normative by the people undergoing them, and what transitions are viewed and experienced as discontinuities and "turning points"? Individuals or families may experience turning points under the impact of internal family crises, such as the premature death of a close family member, illness or physical handicap, loss or damage of property or loss of a job. Other turning points may be externally induced by historical circumstances or events, such as depressions or wars. For example, the Great Depression and World War II caused critical turning points in the lives of the people who experienced them, both in the United States and in Japan.

This emphasis on the importance of external events in inducing turning points should not be misconstrued, however, as cohort determinism. Cohorts encountering the same historical events do not necessarily experience their impact uniformly. Even within a single cohort one would expect the experience of turning points to vary in relation to social background, resources, earlier life history, and personality.[16]

A normative transition may become a turning point under the following conditions:

(1) when it coincides with a crisis or is followed by a crisis, for example, when the birth of a child coincides with the father's loss of a job;
(2) when it is accompanied by familial conflict resulting from asynchrony between individual and collective transitions, for example, when a daughter's timing of marriage clashes with her parents' need for continued support in old age;
(3) when it is "off time," for example, when a father retires early, or when a woman becomes a parent for the first time in her forties or, in a more critical case, in early widowhood;
(4) when it is followed by negative consequences unforeseen at the time of the transition, for example, when a marriage ends in divorce;
(5) when it requires unusual social adjustments, for example, when leaving home involves migration from a rural to an urban area.

This paper examines the themes outlined above through two cohorts in a New England community, whose lives span the period from the turn of the century to the present. By comparing the cohorts' experiences of timing and their perceptions of timeliness we will gain insight into the changes that have occurred in individual and family timing over the current century. In particular we will note differences in the patterns of timing, especially in the transitions to adult roles, and differences in people's perceptions of continuities and discontinuities in their life course.

Cohort Differences in the Timing of Transitions and Perceptions of Turning Points

The Cohorts

The two cohorts from Manchester, New Hampshire, encompass two related generations: the parents' generation (Cohort I), on whom my study *Family Time and Industrial Time* (1982) is based, and the children's generation (Cohort II). For analytical purposes we divided the children's generation into two cohorts in relation to the historical events they encountered as they reached adulthood: those aged 60 to 70 (Cohort II-a) came of age during the Great Depression and those aged 50 to 60 (Cohort II-b) came of age during World War II. While the "historical" (parental) cohort consists primarily of immigrant workers in a textile factory, the children's cohorts consist predominantly of people born in the United States who had made their transition into middle-class occupations and life-styles.

Manchester Cohort I (the historical cohort) consists of men and women who were born before 1900. Most of the members of this group were immigrants who had come to work in what was at the beginning of the century the world's largest textile mill (the Amoskeag Mills in Manchester). They spent the early part of their work lives during periods of labor shortage in the peak of the Amoskeag Company's expansion and activity. Following World War I they experienced frequent discontinuities in their careers. Finally, the Great Depression and the shutdown of the Amoskeag textile mills imposed traumatic discontinuities on their lives when they were middle-aged or older, just at the time when their children were about to reach independent adulthood.[17]

In the children's generation, the older group (Cohort II-a, born between 1910 and 1919) came of age during the Great Depression, a period of economic deprivation and unemployment, while the younger group (Cohort II-b, born between 1920 and 1929) came of age during

World War II, a period of economic recovery and relative prosperity. The two cohorts in the children's generation thus experienced different social and economic conditions in their transition to adulthood. Since the historical cohort had large numbers of children, siblings often belonged to different cohorts.[18]

The Data Source

Data on the Manchester cohorts in this paper were based on a longitudinal historical data set that I had constructed for the parents' cohort for the study resulting in *Family Time and Industrial Time*. In order to compare the cohorts and study change over time, I linked this historical data with a new data set on the children's cohorts, which we generated from intensive interviews, demographic histories, work histories, and migration histories of the children's cohort during the period 1979 to 1983.

By tracing the children of the historical cohort, their spouses, the siblings of the spouses, and other family members in Manchester as well as other parts of the United States, we reconstructed kinship networks as far possible. Following a "snowball" method, we interviewed all kin who responded. We interviewed each person three times, in two- to three-hour sessions with open-ended interviews. The interviews included a broad array of questions pertaining to the interviewees' life histories, work histories, and family histories, their relations with their kin and the older generations, and the support networks available to them in the later years of life, as well as specific questions about timing (being "early," "late," or "on time"). Interview questions also emphasized the individuals' own perceptions of continuities and discontinuities over their life course. We elicited responses about turning points in two ways: by asking directly what crises or turning points individuals had experienced, and by identifying the existence of turning points in the individuals' own references to turning points in their responses.

In addition to conducting interviews, we constructed a demographic history, migration history, work history, and family history for each individual. We then linked this sequential information into a "time-life line" for each individual, reconstructing the individual's life history chronologically, in relation to age and historical time. The time-life line enabled us to compare the life trajectories of the various individuals in the sample; to examine in each individual the synchronization of work-life transitions with family ones; and to relate patterns of timing to the subjective accounts of the life course in the interviews.

The Japanese data, to which this paper will make comparative references, are based on a study of three cohorts in the industrial town of Shizuoka conducted between 1981 and 1984, as part of a comparison of the life course in Japan and the United States. The Shizuoka cohorts consist of Cohort I, born between 1920 and 1922; Cohort II, born between 1926 and 1929; and Cohort III, born between 1935 and 1937. Members of Shizuoka Cohort I are age contemporaries of Manchester's children's cohort (Cohort II, especially Manchester's Cohort II-b). As will be shown below, however, in the reality of its historical experience, Shizuoka's Cohort I more commonly shared a life experience with Manchester's historical cohort (Cohort I). The Shizuoka sample includes only men, while the Manchester cohorts include men and women. Because of the absence of women from the Shizuoka sample, this paper compares the life course patterns of the men only.[19]

Cohort Differences in Timing and Perceptions of Turning Points

In the older cohorts, both in Japan and in the United States, the interviewees attributed almost one half of their turning points to their occupational careers, these turning points having been induced by external economic factors, such as business cycles, the war, and the Depression. The younger cohorts, on the other hand, attributed greater significance to the impact of internal events and to family conditions in causing turning points. Both the historical cohort and the children's cohorts perceived turning points as having occurred in their work life at various points over their entire life course, rather than only early in life. The Manchester historical cohort (I) generally did not identify turning points related to family life, while the children's cohorts, especially the younger ones, claimed to have experienced family-related turning points.

The Manchester historical cohort placed greater emphasis on migration and economic crises as causes of turning points than on personal and familial factors. Indeed the immigrant and working-class background of this cohort led its members to emphasize economic constraints and family needs as being more critical in defining their life course in general than normatively timed transitions. Over half of this group considered the age at which a life transition should occur as less critical than the family's timing of migration or changes in family life dictated by external economic conditions. In Manchester, the older cohort considered interchangeable bouts of employment and unemployment or frequent job changes as the "normal" and expected life-course pattern. The older cohort did not view sporadic periods of unemployment as major turning points. On the other hand, the younger cohorts, especially those who came of age during World War II, viewed periods of unemployment as crises. A linear, continuous work career was rarely part of the reality in working-class life. Until World War II "disorderly careers"—in which people experienced frequent discontinuities over their work lives—were accepted as the norm.

Although economic crises and business cycles affected the parents' cohort in Manchester uniformly, these same external events affected the younger cohorts differentially. The older children's cohort (II-a) encountered the Great Depression when they were about to launch their careers. Unemployment and economic deprivation were, therefore, closely tied to their transitions to adult work-life. The majority of this cohort never completed their high-school education. They had to wander in search of work, and did not achieve stable employment until after World War II. Even after the postwar recovery, their early careers were erratic and subject to frequent interruptions.

The members of this particular cohort (II-a) were coached by their immigrant working-class parents to enter middle-class careers. But the children were unable to fulfill the life plans that their parents had drawn up for them, because of the ways in which the Great Depression and the shutdown of the mills had disrupted their careers. For this cohort the fulfillment of the script of the American Dream, namely, the transition of the mill workers' children into a middle-class life style, had to be postponed until later in their lives, and in many cases, until the next generation.[20]

The younger Manchester children's cohort (II-b), on the other hand, launched their careers during and after World War II. Although the war caused delays and disruptions in the careers of the men who went into the service, it also had a positive impact in the long run, both on the men who stayed home and on those who went into the service. For those who did not enlist, war industries brought employment and eventually recovery from the Depression. For those who joined the military the war opened new opportunities for acquiring skills and competence with which to launch careers.[21]

In trying to account for the experience of turning points in the lives of these cohorts, it is important to explore whether the interviewees' subjective experience of turning points was in some ways related to the timing of transitions. The timing of the three main transitions to adulthood—achievement of economic independence, marriage, and establishment of household headship—was related to the experience of turning points.

For Manchester's historical cohort (I) it was the Great Depression that caused delays in the achievement of economic independence and the establishment of independent households.

In Manchester's children's cohorts, as in Shizuoka's oldest cohort, the delays in the children's launching of stable work careers and establishment of independent families rendered the parents more dependent on their children as the parents approached old age. In Manchester, for both the older and the younger cohorts, the timing of life transitions and the launching of careers were shaped by external economic and social constraints imposed by the Depression and World War II. Not only did these events affect the timing of work-life and family transitions at the point of their lives when people first encountered them; they also indirectly shaped the subsequent flow of careers and family patterns over their life course. The stage in their lives (early in their careers, in mid-career, or later) at which members of these different cohorts encountered the Great Depression and World War II influenced significantly the impact that these events had on their subsequent life transitions and on their perceptions of turning points.

Cohort Differences in Perceptions of Continuities and Discontinuities over the Life Course

While turning points are subjective roadmarks along the life course, formal stages of life are culturally defined and are often institutionalized and legally established. It is important, therefore, to examine differences between the cohorts in the individuals' consciousness of culturally defined continuities or discontinuities in their life course. In Manchester, we found considerable variation among the cohorts in their perceptions of continuity in their life course.

Members of Manchester's historical cohort (I) perceived their life as a continuous whole, rather than as segmented into externally defined stages. They expressed little consciousness of having gone through stages such as "adolescence" or "middle age" or the "empty nest." Nor did they tend to view normative life transitions as turning points as much as the children's cohorts did. Most of the members of the historical cohort (I) considered external events, especially those connected with business cycles and migration, to be much more critical changes than individual transitions. They viewed the declining textile industry, the strike, and the shutdown of the Amoskeag mills as more critical than turning points in their individual lives.

The children's cohorts in Manchester (II), on the other hand, were much more conscious of the existence of culturally defined stages in their life course. Unlike the parent cohort, the children's cohorts (especially the younger one) had a clearer view of their life course as being structured around a sequence of normative transitions and as punctuated by turning points. When asked what had been the major crises in their lives, many interviewees in the younger children's cohort (II-b) mentioned "adolescence" and "middle-age crisis."

The difference between the parents and the older children's cohort and between the younger children's cohort in their perceptions of continuities and discontinuities in their life course could be a result of the stage of life at which they were interviewed, as well as of their location in historical time. Accordingly, the older cohorts may have viewed their life course as a continuum because they were interviewed in their old age, while the younger cohorts were still in middle age or in earlier old age at the time of the interview. The question of the relative significance of life stage during the interview versus cohort experience in the subjective interpretation of the life course represents an important methodological issue, which requires further research. In this case, the "cohort effect," namely, the specific historical conditions encountered by each cohort at various historical moments, may offer a more plausible explanation for the differences in the point of view of the younger and the older cohorts than the life-stage explanation.

Comparisons of the Manchester Patterns with Those of Japan

A preliminary comparison of life-course patterns among the Manchester cohorts with their age counterparts in Japan has revealed considerable differences among cohorts in each country, as well as similarities that cut across both societies. Both the Shizuoka and Manchester cohorts experienced turning points in relation to historical events—the Great Depression and World War II. In both societies this experience of turning points was connected to off-timing as a result of these historical events.[22] While cultural differences in the timing of life transitions and the subjective construction of the life course are significant, the common experience of cohorts in response to shared historical events transcends cultural differences.

In each of the two communities the older cohorts were more likely to experience turning points in their work life, while the younger cohorts had a greater propensity to experience turning points in personal life and family life. In both communities the location of a cohort in historical time was the most significant factor in defining the experience of turning points. Both in Shizuoka and in Manchester the life stage at which a cohort encountered such external events was most significant in causing turning points, as well as in determining an individual's ability to carry out "course correction" (i.e., to take the necessary steps for adapting their life course to these changes). Even though Shizuoka's oldest cohort were age contemporaries of Manchester's children's cohort, their patterns of timing and perceptions of turning points parallel those of Manchester's historical cohort.

The similarities as well as the differences in the experience and attitudes of the Japanese and American cohorts can be explained from a historical-cultural perspective. Both societies idealize linear, continuous work careers and uphold occupational advancement as the ideal script. Both societies introduced mandatory retirement during the early part of this century, thus imposing institutionalized discontinuities on the work life. The surface similarities between the two societies should not, however, obscure the cultural differences affecting the timing of life transitions and the social and subjective construction of the life course.

Historical and Comparative Implications

The differences between the Manchester cohorts' perceptions of timing reflect the historical changes that have affected the life course in American society since the turn of the century, especially the demographic changes in the timing of life transitions and in the synchronization of individual time schedules with the collective timetables of the family unit.[23]

The younger cohorts' consciousness of discontinuities and stages in their life course may reflect the societal recognition of life stages as well as more marked discontinuities in family life, such as the empty nest. In the United States, the empty nest had become sufficiently widespread to account for the younger cohorts' perception of specific stages in family life, especially in the later years of life. On the other hand, the older cohort's perception of the absence of marked stages in their life course is consistent with the demographic realities of the first two decades of this century—a time when Manchester's oldest cohort (I) were immersed in their parenting roles. Typically in this cohort, the youngest child (usually a daughter) remained in the parental household even if he or she was old enough to leave home. It was commonly the youngest daughter who delayed marriage or gave it up altogether in order to remain in the parental home and support aging parents.[24]

The younger cohorts' perceptions of discontinuity in the life course may reflect increasing societal recognition of life stages. The historical trend in the timing of life transitions in American society has thus become more regulated in accordance with socially accepted age norms. The very notion of embarking on a new stage of life and the implications of movement from one stage to the next have become more firmly established. Both in the United States and in Japan, the establishment of legal limits for school attendance, child-labor legislation, and formal retirement all served as official landmarks for such societal and legislative supports for these life stages.

The social and cultural attributes of life stages have by no means been subject to universal definition across time and across cultures even within the same time period. Since the turn of the century, the definition of life stages in American society and their appropriate developmental tasks have become progressively more related to age norms. Thus, the private as well as public consciousness of normatively marked life-course transitions was a much more prominent factor in the earlier life history of the children's cohorts in Manchester. It was, however, less clearly formulated in the counterpart Japanese cohort.

The historical changes affecting the life course were most clearly expressed in the timing of transitions to adulthood. In American society over the past century the timing of these transitions has become increasingly more age-related.[25] Since World War II, transitions have become more rapidly timed and are following a more orderly sequence. Prior to the beginning of this century, by contrast, transitions from the parental home into marriage, to household headship, and to parenthood occurred more gradually and were timed less rigidly. The time range necessary for a cohort to accomplish such transitions was wider, and the sequence in which transitions followed was more flexible and erratic. The increasing age-uniformity in the timing of transitions reflects a long-term historical process leading to the individualization of life transitions: age norms have become more crucial in dictating timing than the needs of one's family or kin.

A similar pattern has also emerged for the timing of later life transitions—into the empty nest and out of household headship. The timing of these later transitions has always been more erratic than that of early ones. Even after the timing of transitions to adulthood had become more compressed, the latter transitions continued to follow a considerable age spread.[26] Eventually, the later life transitions have also become more age related.

Although in the United States orderly and age-related sequencing of transitions is a recent phenomenon, in Japan the strict timing of life transitions and sequencing in relation to age norms has been the modal, historical pattern. In Japan even when familial considerations predominated, the timing of individual life transitions adhered to strict time schedules and sequencing, which had been dictated separately by each local community. Marriage and the birth of the first child as well as the succession to household headship were more uniformly timed. Similarly, the departure of the second and third child from the parental home was synchronized with the marriage of the heir. Hence there existed a delicate balance in the age norms of each local community and family.[27]

In both societies, family considerations and needs had initially taken priority over the individual timing of life transitions. In Japanese society, however, family considerations of timing were traditionally consistent with age-stratified norms governing the timing of life transitions; in the United States, age-related timing of transitions emerged at the expense of familial control. While in the United States timing dictated by family needs was at odds with timing in relation to age norms, in Japan the two were mutually reinforcing.

PART V

Time, Narrative, and Cultural Contact

Frogs Alive, Birds Dead, and Time to Tell a Story: Time, Narrative, and Cultural Context

JOHANNES FABIAN

Of Dogs Alive, Birds Dead, and Time to Tell a Story

JOHANNES FABIAN

Time and the Work of Anthropology

I SHOULD LIKE TO START BY making a distinction. In a general way, "construction of time" has been on my mind for years. The more I think about it, though, the more useful I find it to narrow my focus by looking at constructions *with* or *through* time. "Construction *of* time," I take it, stands for an approach that opposes construed ideas or manners of speaking about (or otherwise symbolically expressing) experience of time to something out of which such constructions are made. Presumably this would be "real" or "physical" time. As an anthropologist I cannot help but recognize in this view the nature-culture opposition that pervades our discipline, serving as a kind of ontological foundation. Talk of "construction" emphasizes, above all, *construed-ness,* that is, a property thought to be the formal prerequisite for there being different, specific types of time, chronotypes. If, to rehearse an ancient argument one more time, experience of time came "naturally," then, nature being one, experience would be one (at least formally; material differences due to conditions affecting experience could still be considered).

I have had more than one occasion to learn that anthropological contributions to thought about time are expected to address precisely this aspect of the question. Our assigned role often is to be purveyors of "comparative" difference. But we cannot deliver different experiences; we can only give accounts of accounts of different experiences. Furthermore, constructions of time cannot be accounted for (or recounted) without construing time. There is no way of thinking about chronotypes but chronotypically. Unless we want to accept an infinite regress into relativities, we have, in my view, but one direction in which to move: We must seek a better understanding of a praxis, ours as well as theirs, of giving accounts of difference (and, of course, of sameness, and of "movements" between difference and sameness) that are constructed "temporally," *with* conceptualized experiences of time.

What, then, can I reasonably offer as an "anthropological" contribution to a debate about chronotypes? Anything that would just be an ethnographic case-illustration is excluded; I never studied "conceptions of time" in another culture. What I have been interested in is the role of time—time passing, time experienced, time conceptualized—in the production and communication of anthropological knowledge.[1] Up until now I have concentrated on a critique of what appeared to me an all-pervading trait in anthropological talk and writing. I call it "temporal distancing" and the discourse built on such distancing "allochronic." It began with the realization that the practice of anthropology[2] has been characterized by contradictory "uses" of time. What we call "fieldwork"—research that has become institutionalized as our principal source of empirical knowledge and, at the same time, the principal *rite de passage* toward professional recognition—is carried out among people who are our contemporaries. Apart from observing, measuring, depicting them, we must communicate with them if we want to learn something about them. Successful communication demands that we share time, that we are coeval. This involves more than mere synchronization of separate courses of action. It requires participation in events that are experienced and recognized as such by the participants (the idea of "speech events" has been central in communicative approaches to ethnographic research). A contradiction arises when we then turn around, as it were, and deny the people whom we investigated coevalness by pronouncing on them an allochronic discourse. This is what happens: When written about (or otherwise represented according to conventions of scientific literacy), anthropology's object, the Other, has consistently been placed in a time other (usually earlier) than that in which the writing anthropologist places himself or herself.

Of course, the qualifier "allochronic" takes its specific meaning against the background of a general "chronicity" of discourse. If we want to get from verdict to project, and from diagnosis to therapy, we must try to understand more about discursive chronicity in general. I can think of two roads to such a goal: One is to examine "theories of time," which are offered in abundance; to weigh and select and perhaps to adopt what appears valuable. In a modest way, I tried this kind of roaming and ruminating approach when I prepared myself for writing *Time and the Other*. But this is a once-in-a-lifetime luxury for a practicing anthropologist, and, while I try to keep an eye on the literature, I cannot claim comprehensive knowledge of the subject.[3] That leaves me with the other possibility, which is to build small theoretical constructs, either from illuminating episodes of research[4] or from ordinary "chronic" introspection/experience (as philosophers do), and to apply their interpretations to our shared discourse of "ethnography" or anthropology. In either case, the aim would be to understand what *kind* of discourse is being produced with what kind of *chronic* devices. In this essay I shall take the introspective approach. I will narrate, and then reflect on, two experiences that, I believe, enable me to communicate some insights about constructions with time in anthropology without burdening my account with ethnographic detail.

Nevertheless, before I get there I can offer at least one example of how the problem of temporality and discourse may be encountered concretely in the course of ethnographic research. I have just completed a study of conceptions of power based largely on work with a group of popular actors in the Shaba region of Zaire.[5] The most important corpus of data consists in this case of tape recordings made during rehearsals and performance of a play constructed around a proverblike pronouncement on power. Methodologically, any ethnography based on documentation of this sort must address the question of relationships between event and discourse, in this case between performance and text. In what sense can ethnographic texts—here taken as the transcripts and translation of recordings made while participating in events—be taken as documents? What happens between the event that takes place and the text we take away from it? These questions arise routinely in the course of ethnographic work and have been dealt with by some anthropologists with considerable care and sophistication.[6] In this specific case I was, however, troubled by a most vexing problem. Access to events for the purpose of recording was easier in this project than in any I had carried out before; the actors welcomed my presence at all times. Generating an adequate text of the play also seemed a less difficult goal to attain than, say, preparing usable texts from interviews at a workplace or from the group meetings of charismatics. Paradoxically, it turned out that the nature of the events was in this case such that my recordings only produced fragmentary texts and, at any rate, never a definitive "script" of the play. When I tried to understand what was happening it occurred to me that the solution to the problem of the "missing text" must be sought by considering different ways in which process- and event-immanent time shapes the textual record of performative events. In a nutshell, I found that over the course of the play's production and performance the relationship, with regard to "timing," between discourse/talking and performance/acting became inverted: As the process of constructing the play moved from exploratory discussions to rehearsal of dialogue to actual performance, timing—timed action, only some of which may be verbal—became more important than talking. Consequently, the closer the process gets to actual performance the less "talk" there is to record, transcribe, and make into "texts." Even the talking that can still be recorded is now all acting. It is delivered in ways that resemble gestures and movements more than statements and propositions. What the tape recorder catches and the ethnographer manages to transcribe is, in some extreme instances, literally but a string of timed sounds whose "meaning" lies in the action they perform. By implication this means that the more actively and intensely the ethnographer participates in events (which are cultural performances) the less likely he is to carry away the kind of textual documentation—discursive, informative accounts—that seems to be required for good ethnography. This was not a ground-shaking discovery, to be sure, but it has helped me, first, to overcome what I call textual fundamentalism—the notion that the more text we take away from an event the better it is ethnographically documented—and, second, to shift attention from the semiotics of time to the pragmatics of temporality (from what time "means" to what we do with time).

Quite possibly, this "ethnographic example" raises more questions about time and anthropological research than it answers. One insight I derived from it—that we must consider "timing" as well as time—I shall take up again in the last section of this essay; but first I want to follow an introspective line of reflection.

Of Dogs Alive . . .

Here, then, is the first story. Billie, our dog, is a West Highland White terrier of, I believe, striking and distinguished appearance. One day, my friend and I took her for a walk in the streets of Amsterdam. Billie was very young then and we kept her on one of those long leashes that come with a self-rewinding reel. We were walking and talking and did not pay much attention to Billie. We "knew" she was trotting along happily. Suddenly my friend, who had been holding the leash, stopped. She turned around to look back. "Look, Johannes," she said, "there is a dog exactly like ours." I looked and saw the dog. It took us a split second to realize that it was Billie, at the far end of her ten yards of freedom.

Was what occurred here just a cognitive lapse (a *Fehlleistung* of the kind that has been recognized by psychologists as revealing, if predictable)? If it was, it will be for experts to provide an explanation. Here is my understanding of what actually happened. All three of us were taking a walk. We were engaged in action, participating in an event, something that requires contemporaneous togetherness. Since my friend and I were walking closely together we expected our dog to be close to us. When we suddenly saw her ten yards away we corrected or changed her identity—look, there is a dog just like ours—rather than revising our assumptions or expectations regarding the event. The latter would have required us to consider that Billie might have moved away from us and thus not have been where we expected her to be at the time when we perceived her elsewhere (elsewhere, that is, than expected). Rather than experiencing our dog—and, through her, us and the event—as being there as well as here, we saw another dog.

I believe that the story can help us clarify what happens in allochronic anthropological discourse. Let us say that the "problem" raised by the incident is that of the perception of identity and difference. This, to me, seems obvious enough not to need further reasoning at the moment. It is more important to consider the situation in which the problem arises. Identity—sameness and otherness, as the case may be—is predicated at a specific moment and from a concrete position. Given such a historically contingent standpoint, sameness, it would seem, requires or, if that is too strong a word, goes together with presence *here,* with sharing the same time as well as the same place as the one who considers identity. "Same time" is of course more than physical synchronicity; it is relational, shared, bounded by what I called the "event." Nor is "same place" a physical fact. It is a construct of expectation (and memory, if it is to remain the same for any length of time). On scales larger, but not much larger, than that of the personal experience I reported, shared time and space are fused into identities we call community, society, civilization, and history. So strong are the fusions of *here* and *now* that to encounter human beings *there* seems to make denial of sameness inevitable. Identity, in the sense of identification with another, is broken, changed into otherness or, to be faithful to our story, into likeness, which is of course the only form in which otherness can be experienced short of total otherness or transcendence.[7]

The point of our story is that, normally, otherness is experienced only *as if* it were a fact; the illusion is quickly corrected. It took us just an instant to realize that Billie was there, not a dog like her. (That she was still on the leash, incidentally, may have helped us with the correction, but I don't think it affects the argument. The leash only gave concrete symbolic expression to our being "bound" together in the same event.) So why, when we produce anthropological discourse about contemporaneous peoples, do we persist in maintaining analogous illusions of otherness *as a fact*? Why can't we admit that the savage, primitive, underdeveloped, the Nuer, Eskimo, Bororo, Aranda, and all the other Others are our *compagnons de route,* fellow travelers (in a concrete, historical way, not an abstract, philosophical way)? Why will we, rather than adjusting our definitions of time-space togetherness, posit a radical sort of otherness, which we then disguise with all that talk about comparative appreciation of human diversity? An idea worth pursuing in this connection might be the following: Could it be that the deeper significance of the famous "comparative method" that became a powerful and unifying paradigm in the life sciences and social sciences has been a kind of secularization of conceptions of religious and transcendental "otherness"? In his recent account of the discovery of geological time, S. J. Gould points to some of the religious-mystical aspects of this story.[8] He also addresses the problem of systemic ("Time's Cycle") and narrative ("Time's Arrow") conceptions. He seems to plead for a sort of peaceful coexistence of the two metaphors, on the grounds that they are metaphors. I do not think that this is a satisfying solution, at least as far as cultural evolutionism is concerned. But more about that later.

The question before us is: Why must distance be difference; why will we, where difference is not obvious, create distance? Because of a cognitive "lapse"? This, although such a lapse made us think about the question, will hardly be acceptable as an answer. It takes purpose and effort to impose otherness and to make imposed and therefore illusionary otherness determine our relations with other peoples. The otherness anthropology delivers for *our* purposes—for visions of progress and evolution, for projects of colonization, for schemes of development—is never of a disinterested, objective sort.

But here I do not want to pursue a line of thought that must bring us too quickly to the role of interests and power in anthropological discourse. The point I want to make is of a different kind: Constructions of otherness do not begin with evolutionism and other schemes of distancing whose ideological character we now recognize; they are already built into our very presentations of identity/sameness as an *exclusive* "here and now," which we accept without much questioning. One trick especially, which we seem to play again and again—the trick of denying coevalness, same time, to those whom we perceive as distant and different—works as a "construction with time."

The attentive reader will have noticed that my reflections so far do not cover all the logical possibilities of what I called space-time fusions. We have linked identity/sameness to "here and now" and difference/otherness to "there and now." What about the remaining possible combinations, "there and then" and "here and then"? It would seem that these concern above all archeologists and historians. Their ways of construing interpretative discourse (of telling stories) are, to say the least, similar to the anthropologists'. Therefore, "allochronism" may have to be examined critically even in disciplines that find their object in times other than ours. It may boil down to this question: If one realizes that temporal relations to a scientific object are *construed* rather than given, what difference does it make if the "material" for construction is found in the historical past or in the present?

Of Birds Dead . . .

The present focus of critical thought in anthropology and elsewhere on constructions of otherness is, I believe, salutary, and I hope to have given some indication why it bears on a discussion about chronotypes. However, the unintended effect of a strong focus can be that concentration is mistaken for exhaustiveness. Critique of allochronism is important, but there is more in anthropological discourse and practice that can make us think about "construction with time." To be sure, concepts such as identity, likeness, sameness, and otherness are foundational; yet they mark points of departure that by and large remain outside the daily business of ethnography. Usually, anthropology is not concerned with ultimate foundations of knowledge; it tries to give specific explanations and interpretations, that is, to provide alien human conduct or its products with "meaning." I now want to argue that the production of meaning, too, builds on constructions with time. And I will introduce my argument again in a roundabout way.

Here is the second story. Not long ago, I made a trip to Belgium. I was driving through the pastures of Holland, past windmills, across rivers that carried barges toward ports on the sea. The weather was fine, traffic was light, and I was in a pleasant mood of anticipation. I would meet friends in Brussels and get away from the chores of teaching and administration to do some work in colonial archives, which always promise interesting discoveries. Suddenly a blackbird crossed my path. It was flying low; at the speed I was going it was impossible to brake or make an evasive move. I heard the thud as the bird hit the grille. In the rearview mirror I saw its fluttering body come to rest on the highway. I was shocked; my good mood disappeared and I knew at that very instant that things would go wrong for me. They did. In Brussels I had car trouble, and several annoying and sad incidents happened in the weeks that followed.

One need not be a specialist—psychologist, folklorist, or some other kind of hermeneut—to understand what this story is about. We probably all share a folk-category that helps us to recognize the genre: it is about a bad omen. In fact, if the identity of the storyteller would not make this unlikely, we could file the tale away under the rubric "Well-known superstitions." But I am not a superstitious person; I have killed birds before while driving, without that feeling of foreboding. Nor are blackbirds flying low on my cultural list of bad omens; black cats crossing from the left are.

In other words, I think we can dispose of some of the easy ways of assigning meaning to this story. Instead I shall try to develop an interpretation that will make it clear why I think it is, in the end, about construction with time.

To begin with, omens do not become omens by being perceived (what is being perceived is an event or occurrence). Omens become omens only by being told. "Being told" requires minimally that the omen be related (*relatio,* an account of . . .) to an event actually experienced. (Potential omens are not omens; they are just symptoms of pessimism.) This sort of narrative construction of meaning is easily enough understood in cases where we "remember" omens, as it were, *post factum,* when they have "come true." A past event takes its meaning (or, at any rate, additional meaning) from being experienced later as fulfilled, as giving to the ending of a story a quality we call meaningfulness. But strictly speaking omens are omens right away. They don't have to wait for confirmation by future events. What happens when we construe events as omens at the moment when they occur? I can think of only one answer: Events acquire meaning as omens when we construct them as a past for narratives to build on. That we construct that past in the form of a future (they are omens, after all) is, in my view, no objection to the thesis that *in ominous experiences contemporaneous events are being related to the person who experiences them as essentially past* (with all that this may involve: for instance, the feeling of inevitability we get from omens may be nothing else but recognition of the fact that the past cannot be changed).

So, an event that is present (for all practical purposes, because we consider the present as experienced, not as the unextended moment of duration) is construed as past. This is a necessary condition for it to become an omen but it is not sufficient. We have hinted at the fact that in a given culture there may be categories, hence criteria, that impose ominous meaning on events through classification. But such a "cognitive" interpretation only satisfies those who posit that all cultural meaning is just a matter of classification. After having classified omens as instances of cultural classification we are still left with a residue, as it were, namely the fact that ominous events need to be distinctive, strange, and probably also rare, sudden, and fleeting. A certain symmetry seems to be required between the omen and its fulfillment in that the latter, too, is an event out of the ordinary, a misfortune or a case of good luck, as it may be.

But let us return to the story of the dead bird and to the temporal conceptualizations we detected through it. I think it can help us to take a fresh look at one of anthropology's established discursive practices, that of incorporating strange, disquieting cultures and, indeed, our traumatic confrontations with these cultures (such as discovery, conquest, colonization) into *narratives of cultural evolution,* that is, narratives of fulfillment. To this view it could be objected that evolutionists have been concerned with origins, beginnings, not with endings. The origins of religion, law, the family, the state, agriculture—those were the great topics of evolutionary anthropology. True, but historians of evolutionism in anthropology showed a few decades ago that these topics were selected and treated such that they served to explain, and often legitimize, the modern state of affairs. The important thing in tales of evolution remains their ending.[9]

That was the case in the nineteenth century. More recently stories of cultural evolution have been offered as accounts of law-governed changes. Evidence for change and process, it is said, comes from data that document cultural differences. Evolutionary sequences are established by the disinterested operation of a taxonomic method called "comparative," one of whose attractions has been that it could serve as the great equalizer and routinizer of human history; nothing is extraordinary, nothing strange or irrational, nothing catastrophic or ominous before the impartial theorist who judges everything according to transhistorical values such as function, adaptive value, systemic integration, and so forth.

But there are reasons to scrutinize this received view. First, it has been a basic assumption of the comparative method that societies radically different from ours, even though they exist contemporaneously with us,[10] "represent" the past, to say the least, or that they are (the) past, to say the most. Second, contrary to the image of scientific equanimity that evolutionary theorists like to project, there has always been a practice of selecting poignant cases and striking *topoi* (literally and literarily) for the construction of developmental sequences and stages. Remember the fascination that "survivals" had for nineteenth-century evolutionists. These cultural oddities, although found in the present, really were thought of as cracks through which the past creeps up on modern society. This is aptly illustrated by the following passage from one of the early introductions to anthropology intended for the general public:

> Customs may linger on indefinitely, after losing, through one cause or another, their place amongst the vital interests of the community. They are, or at any rate seem, harmless; their function is spent. Hence, whilst perhaps the humbler folk still take them more or less seriously, the leaders of society are not at pains to suppress them. Nor would they always find it easy to do so. Something of the primeval man lurks in us all; and these "survivals," as they are termed by the anthropologist, may often in large part correspond to impulses that are by no means dead in us but rather sleep; and are hence liable to be reawakened, if the environment happens to supply the appropriate stimulus. Witness the fact that survivals, especially when the whirligig of social change brings the uneducated temporarily to the fore, have a way of blossoming forth into revivals; and the state may in consequence have to undergo something equivalent to an operation for appendicitis. The study of so-called survivals, therefore, is a most important branch of anthropology. . . . It would seem to coincide with the central interest of what is known as folklore. . . . Firstly, the survivals of custom amongst advanced nations, such as the ancient Greeks or the modern British, are to be interpreted mainly by comparison with the similar institution still flourishing amongst ruder peoples. Secondly, all these ruder peoples themselves have their survivals, too. Their customs fall as it were into two layers. On top is the live part of the fire. Underneath are smoldering ashes, which, though dying out on the whole, are yet liable here and there to rekindle into flame.[11]

Incidentally, although nineteenth-century evolutionists still acknowledged something of the threat that survivals pose, and hence their presence, latter-day neo-evolutionists who, following the lead of Radcliffe-Brown[12] and others, subscribe to the idea of "functional unity of cultural systems" offer glib solutions for the problem:

> In the nineteenth century many anthropologists glibly spoke of *survivals,* traits that at one time were functional but that with the alteration brought on by cultural evolution had lost their utility. After considerable controversy it has come to be recognized that many cultural traits lose their original function and meaning, but few indeed can become utterly useless. . . . Among the favorite candidates for a truly useless, vestigial survival are the sleeve buttons on a man's suit jacket, which were formerly used to keep shirt ruffles out of

the inkwell. Even here, however, four extra buttons per jacket are not exactly useless from the point of view of the button manufacturer.[13]

What I am leading up to is the proposition that much of cultural evolutionism has been like the story of the dead bird writ large: Certain other cultures, or certain traits in our own culture, become omens when, although we experience them in the present, we construct them as past because their striking otherness confers (additional) meaning on "civilization," on its perils as well as its precarious achievements. Now, to find and interpret omens is the business of divination, and if there is one candidate for the most divinatory approach in the social sciences it would be, I am arguing, the theory of cultural evolution. Not, as I shall explain presently, because it is guesswork (although much of it is), but because it makes otherness ominous by construing it as "past future."[14]

I guess it does show that I am less than enthusiastic about cultural evolutionism. This is not the place, however, to discuss my reasons in any detail.[15] As I said, the divinatory approach as such is among its lesser faults, if it is indeed a fault. Natural scientists, anthropologists, sociologists, and historians, have found divination to be beneficial, even necessary, to the production of knowledge. B. Jules-Rosette, for instance, has written on oracular reasoning in the social sciences, and C. Ginzburg sketches a sort of world history of divination, with references to an interesting collection of essays on divination by J. P. Vernant and others.[16] Jules-Rosette defends oracular reasoning as something that folk-inquiry and scientific investigation have in common. Her argument focuses on the "logic" of such reasoning. Ginzburg wants to remind us of the ancient origins and modern revival of "conjectural" knowledge as a legitimate alternative to "scientific" demonstration. Both authors provide a background for my attempt in this essay to show the connection between omens and stories of fulfillment in certain kinds of anthropological discourse, for example, evolutionism. To be able to show such a connection does not of course "disprove" evolutionary findings on any logical grounds; it merely demonstrates the workings of time constructs in evolutionary narratives. The latter may or may not be a reason to object to cultural evolutionism as a doctrine.

Time to Tell a Story

Considerable efforts of thought and inquiry have been spent on narrative and time, or narrativity and temporality. In this brief reflective essay I have no ambitions to add to the philosophical and historical depth or the methodological sophistication of that debate. I do want to make an attempt, however, to relate some thoughts on time and narration to the current critique of anthropology. I have in mind above all the debates on "poetics and politics of ethnography"—to borrow the subtitle of a collection of essays edited by J. Clifford and G. E. Marcus.[17]

Preliminarily, two brief remarks on the target of critique are in order:

First, critique, I believe, must aim above all at anthropological praxis. *Ideologiekritik* is fine, but leads nowhere unless one can show as precisely as possible where and how ideology informs practice. I also think that it is an ideological obfuscation when anthropological praxis is conceived primarily or even exclusively in such abstract terms as observation, description, analysis, interpretation. We finally have come to realize that, when the chips are down, that is, when anthropologists have gone through experience in the rough, when they have noted down, classified, and theorized, *writing* is really what they do. But even that is not concrete, practical, enough. When we write we narrate, we tell a story; and narration is never a "natural" way of transposing experience and insights into prose; narration is construction with time. This is not a view of anthropological praxis with which most of my colleagues would agree. Inasmuch as any thought at all is given to the issue, prevalent opinion probably still holds that anthropological writings ("ethnography") are scientific representations of observed facts and of relationships (causal, functional, symbolic) between facts. Up to a point this is the case, but to say that there is nothing more to it would miss the specific predicament that anthropology shares with history (at least as long as we accept that ethnographic knowledge is directly or indirectly based on "fieldwork"): it has no access to facts except as embedded in events. No amount of quantification or structural formalization, no sophisticated display of graphs and illustrations, can cover the fact that accounts of events must be stories, narratives.

Second, neither do I think that awareness of and experimentation with literary genres in "postmodern" ethnography will accomplish much. For instance, one of the main contentions in current critical debates about anthropological writing is that "modern" anthropologists were (and are) writers who have been using the literary idiom of realism. Presumably, that means that they told, or described, what they experienced "as it is." As far as I can see, if this was ever the case, it was so perhaps with writing under the paradigm of structuralism-functionalism. Ethnographers were expected to produce accounts that were valued by the degree to which they matched presumed reality, rather than by the meaning or aesthetic pleasure they created. Not that there was no meaning to this sort of work;

it was just that meaning had become contracted, as it were, to the fit between reality and ethnography. Isomorphic representation excluded movement both in the societies described and in the descriptions. Social realities were represented in time slices, synchronically; accounts were monographic, if not diagrammatic; and almost everything that became the subject of writing was called a "system." Structuralism-functionalism was heralded as a decisive advance over evolutionist "conjectural history," but the euphoria it generated was brief. Sooner or later, at the latest when attempts were made to synthesize the snapshot ethnographies that have been piling up, anthropologists reverted to telling stories. Evolutionism became modern once again and almost everybody now seems to have rediscovered "historical anthropology." So, it is not so much the denunciation of realism, that is, of one genre of writing, that should preoccupy us; rather we should seek a better understanding of the means and devices most widely employed in the construction of anthropological narrative.

Let me recapitulate the main points that, I think, have come up in these reflections so far. I began with two instances of rather ordinary experience. I tried to show that and how time constructs were involved in shaping, out of quasi-instantaneous occurrences, events of which I can give discursive accounts (as anecdotes, stories, narratives). From the first story—of dogs alive—we took away the insight that a construct I called time-space fusion may influence the perception of identity as sameness or otherness. The second tale—of birds dead—served to illustrate how an event becomes an omen by being construed as past at the very moment it happens. To put this another way: The dog story tells of that which is same in the other being made "other" (Billie becomes a *different dog*); the bird story tells of that which is other in the other being made same (as an omen meaningful *to me*). Both operations are carried out with the help of "constructions with time."

Objections may be raised to the way I set up my narratives, especially the bird story: Am I not using too many props to build up a trivial incident so as to fit my larger argument? The image of a bird conjures up the augur; a black bird evokes bad luck; the small, innocent bird's being hit by a powerful vehicle suggests fateful power relations, perhaps also fateful guilt; the green landscape in fine weather invites intrusion of the bad and ugly. Not all omens need killing and death as "signs"; yet death is essential in this story because it makes the connection between the event experienced in the present and an inevitable past. To all this I can only respond: I reported the event as I experienced it. That its "superadded" meaning somehow was derived from, and thereby confirmed, my critical views of anthropology illustrates the working of constructions with time "rhetorically." But that was the purpose since these reflections are about the rhetorics of otherness.

Temporal constructs of the kind I identified, I then proposed, are employed when anthropological discourse, not all of it but certainly a crucial part, assigns identity and meaning to cultural difference/otherness. Of course, insights derived from reflections on personal experiences cannot simply be projected onto a collective practice. To make what, to me at least, has been evocative a convincing interpretation needed further argument.

Throughout I have worked with the presupposition that empirical ethnographic research is not transformed into anthropological knowledge *ex opere operato,* merely by operating received techniques and methods. It matters little whether an anthropologist opts for a hard, scientific, preferably quantitative approach, or for one or the other kind of interpretive hermeneutics. The former may prefer analytic rigor and distance, the latter poetic imagination and empathy; in the end, both must tell a story. That narration is inescapable is, therefore, the hinge of my argument. This is asserted as being true to the best of my knowledge, but it is of course not proven or demonstrated in these reflections.[18]

If my argument has any merit, it suggests certain specific tasks to be tackled within anthropology and some challenges of a broader philosophical and, in my view, political nature. Among the tasks is to explore in more detail how *identity* and *meaning* are assigned in anthropological/ethnographic narratives with the help of constructions with time. Critical questions to be asked would be: Are the temporal constructs we routinely operate *necessary* for the generation of our stories? In what sense can they be *manipulated* for ideological purposes? Where, in anthropological narratives, are the boundaries between fiction and deception?

To begin with the theme of identity, our stories of cultural difference seem to be told from a position that defines our own identity in terms of a time-space fusion between here and now. I speak of "fusion" because of the point made by the first of my stories: So strong is our inclination to think of identity/sameness as here and now that we seem to be almost incapable of admitting that it may also be found there and now. Is that kind of here and now a natural given? Certainly not; although, to be real, identity/sameness must be anchored in real space and time, it is not a physical fact. I don't think there is an anthropologist who would care to deny that what counts as here and now is culturally determined. So we are at least far enough along to relativize our own position. But we should go further. Examining temporal constructs may reveal that cultural relativism has not prevented us from considering certain fundamental notions, such as sameness and otherness, as natural, necessary and not debatable. As I have argued elsewhere, time-space fusions amount to political cosmologies; sameness and otherness are then not objects but categories of thought: "A discourse employing terms such as primitive, savage (but also tribal, traditional, Third World, or whatever euphemism is current) does not think, or observe, or critically study, the 'primitive'; it thinks, observes, studies *in terms* of the primitive. *Primitive* being, essentially a temporal concept, is a category, not an object, of Western thought."[19]

Of course sameness and otherness are categories of thought, says the philosopher; what could be wrong with that? What is wrong is to substitute categorically imposed otherness for experienced otherness. What is wrong is to see Billie as another dog just because she happens to be outside the time-space fusion we take for granted. What is wrong is to allow categories of otherness that are the product of specific and contingent time-space fusions—the savage, the primitive, the folk, the underdeveloped are such categories—to become the points of departure for and the defining terms of our relationships with other societies and cultures. Some anthropologists always knew and many know it now: Experience of difference and otherness begins only when received time-space fusions begin to come undone. What would have to happen for that insight to become collectively and politically accepted? What would our relations to the "Third World," to the "underdeveloped" and "tribal" societies, to ethnics and minorities be like if they were no longer established categorically? What kind of changes would have to occur in our communities, societies, and nations? All these conceptions of identity as sameness rest on time-space fusions that, although historically contingent, are most of the time kept beyond the reach of critical debates.

So much on time-space constructs and *identity*. The other issue raised in our thoughts about time in anthropological stories was *meaning*. Many of the stories that make up anthropological discourse have been accounts of meaningful fulfillment (or of meaning conceived as fulfillment). Meaning construction here seems to revolve around the alternatives of incorporation (positing something as an omen rather than a meaningless random happening) and exclusion (ignoring the event). In theories of cultural evolution anthropologists have scanned a seemingly endless variety of cultural expressions. Contrary to their own protestations, the schemes of systematic connections and orderly progress they produce are never simply the result of detached, disinterested classification. What is admitted as significant and what not will be determined by a method akin to divination. Only those traits of different cultures are assigned meaning that have the status of omens, that is, of events that find fulfillment in our own civilization. I argued that two requirements must be met by events to become omens: First, they must be striking, curious, or extraordinary for the one who perceives them (that the extraordinary may be routinized in the form of culturally established lists of omens does in my view not affect the argument).[20] Second, such events must be experienced as past in the sense that their relation to the future is not open and undetermined. This is precisely what happens in stories of cultural evolution that are based on that eminently temporal construct called the comparative method: ways of life, modes of thought, methods of survival that exist now, but not here, are related to our own now and here as past. Instead of confronting other ways here and now as challenges to our own ways of life, modes of thought, and methods of survival—something that would require us to acknowledge otherness as present—we incorporate them as omens into our stories of fulfillment. What, then, are the chances for us to establish meaningful relations with other cultures and societies that could be the foundation of just and rational politics if we already start out with a surplus of meaning that determines our very perception of cultural difference?

In a way, the questions raised so far imply their answers. To break up ideologically cemented space-time fusions and to abandon a divinatory approach to cultural difference would seem to be the alternatives to the positions and practices we criticized. But such global projects are not really up for discussion here. What is being questioned is the "time to tell a story." Can we recommend that anthropologists continue to tell their stories, provided they mend their ways with temporal constructs that critique reveals as ideological? Or should we envisage the end of narration as we know it?

Time and Narration: Open Questions

An earlier version of this essay ended with these questions. It was suggested that I might think about some answers. Searching here and there I came upon David Carr's recent attempt to put what I called the inescapability of narration onto new philosophical foundations. That he set out to do this in order to show what happens in history writing (of anthropology he has a rather unhistorical image) takes nothing away from the relevance of his findings for the problems discussed here. His characterization of what I called allochronism will show why I read this book with the greatest expectations:

> . . . the view of the world being expressed here [in the writings of Ricoeur and others on time and narration] is itself a narrative-historical view. The scale from animals to "primitive" societies to "historical" Western society is not merely evaluative but developmental. This is the same conception which led Hegel, in his lectures on the philosophy of history, to treat "China" and "India" as *precursors* of the Western world even though both continued to exist in his own day as they do in ours. Now "primitive" societies, which exist in the present all over the globe, are relegated to the past by being regarded as leftovers from an earlier stage of humanity. What is in fact synchronous is arranged on a diachronic scale. What is more, the latter constitutes a dramatic story, the *Bildungsroman* in which Western man represents the maturity of civilization and the realization of all that is human.[21]

As it turns out, Carr realizes that a Western point of view we recognize as "self-congratulatory and parochial" should make us "cautious about asserting the universality of narrative time,"[22] but that is as far as he is prepared to go (not far enough, if the demands of a critical anthropology are to be met). The reasons I will now give for being disappointed with his conclusions will not produce the answers to the questions about narration that my own treatment left open, but will help recast those questions in terms that make them interesting beyond the confines of anthropology.

First let me summarize my understanding of what Carr wants to accomplish. He enters the debate about narration and narrativity by taking a position against a view he feels is too depressing and cynical to reflect ordinary experience, namely that there is no natural, certainly no necessary, connection between events and the stories we tell of them.[23] According to this view, all stories are impositions on reality and narratives are just one genre among others. Going back to Husserl and Heidegger (and occasionally to A. Schutz) Carr finds that a narrative structure—a sequence, for instance, of beginning, middle, and end—is inherent in ordinary time experience and that it naturally comes forth in ordinary, "pre-thematic" reflection on, and accounts of, experience. This applies to temporality not only in individual experience, but also in collective, social experience in the various kinds of groups in which we find ourselves. His argument is of course much more detailed, and one of the most engaging traits of this study is Carr's eagerness to formulate and discuss objections. In the end, however, he does convince himself that there is a fundamental continuity between the narrative structure of experience, both individual and social, and the narrative accounts given of experience. This may be taken as an indication that he takes a "realist" position[24] against the cynical nominalism of Mink, White, Barthes, and others. But matters are not as simple as that (on either side of the controversy). Carr keeps his position philosophical, that is, open, when he admits that even a successful demonstration of homologies or analogies between individual and social experiences of temporality as narrative raises a crucial question (which is of course also central to anthropology). The question arises when analysis is extended from ordinary accounts to, say, historiography. Convincing oneself that we make our "group" experiences "narratively" (as stories with a purpose attained by actions that have a beginning and an end) is one thing. Determining the validity and relevance of accounts given of that experience is a different matter altogether. It all boils down to the question: Who is "We"?

It comes as no surprise (at least to me, because I share Carr's taste on this point) that he turns to Hegel for a solution. Hegel's Spirit, Carr suggests, should be understood as the vision of a We as subject of universal history. It is also predictable that he finds that solution wanting[25] and that he ends his study by resigning himself to the general affirmation that our Western culture and others who may have very different stories to tell, or none at all, are nevertheless solidary in our common "struggle against temporal chaos, the fear of sequential dispersion and dissolution, the need to kill off Father Time or at least stave off (postpone?) his attempts to devour us all."[26] This may indeed be everybody's ultimate conclusion, but before we get there much remains to be argued for the fact that it does not absolve us from the critical labors of understanding the sort of politics of time that I have tried to show are served by anthropological (and historical) discourse.

By way of conclusion, I want to give at least some indication of how I think the struggle can continue. I shall attempt to do this by addressing the two principal assertions made by Carr.

First is his assertion that narratives are not mere arbitrary impositions on experienced reality. I agree, for if this were the case there could be no wrong narratives, or, to put it more exactly, there would be no way of choosing between different narratives except on purely formal, perhaps aesthetic, grounds. However, the reason to reject this view cannot be that narratives have a natural justification in their being grounded in the natural narrativity of experience. Carr does not see that if this were so, there would be no reason to worry about them (or about historiography, or anthropology). A search for what is wrong with certain narratives would have to be directed to what is wrong with ways of being—perhaps a valid project (think of Marx or, for that matter, Heidegger), but one that can be carried out only through a critique of narratives. Meanwhile, distortions, lies, self-serving ideological misrepresentations can be recognized as such and are not sufficiently explained in terms of (individual) moral weakness. They do serve power and collective, political domination, and since those who pursue such aims act "naturally" (unless we want to entertain the idea that wickedness is the work of a supernatural agency such as the devil), there must always be a "difference" between being/experience and narration that can be exploited in this manner.

Carr's second principal assertion is that in exploring the phenomenon of temporality and hence narrativity in human experience we must move from Husserl's and Heidegger's "phenomenology" to Hegel's. Again, I agree. The problem is the "We"; but we shall not get from insights about individual experience to an understanding of collective narratives by way of "analogy." Hegel did not; his approach was dialectical, a way of thinking for which Carr does not seem to have much use. The price he pays for neglecting dialectics is that he does not notice that both his principal assertions fail to recognize the critical value of the notion of narrativity. That value does not consist in building "natural" bridges between being/experiencing and giving accounts of experience. On the contrary, I would argue, narrativity and narration are *mediations* of both; they are moments in a dialectical process of knowledge production whose nature is both to connect and to separate the two, and to do this not once and forever, but again and again. Only if we take seriously the mediative nature of stories can we do more than end in relativistic resignation or in vacuous contemplation of diversity, as the case may be. One of the things anthropology has accomplished by struggling with the notion of "culture" is to recognize its mediative function. To regard different cultures as different versions of a story must be kept as an open possibility unless we want to opt for outright epistemic imperialism, however mitigated by protestations of tolerance for other views.

One point I have tried to make in this essay is that the problem with time and narration does not only arise when it comes to giving "meaning" to experience; it is involved in the very perception and conceptualization of "identity" (of characters and events) that precedes emplotment in stories. Carr cannot fail to notice this when he discusses Hegel's "dialectic of recognition," but by the time he gets to this point (in the next to last chapter) he has cast his dice for what, I fear, remains an undialectical conception of "Time, Narrative, and History."

One final thought that was adumbrated when I claimed that it is in the nature of narrativity as mediation to require that stories be told again and again: It is not only temporality, the experience of time passing (and of conceptualizing that experience in "chronotypes"), that both enables and limits our understanding of stories; it is also what I have called "timing," a way we have with time when we produce narratives/texts, as shown in my one example from ethnography. Again, Carr seems to be aware of this when he contrasts Husserl's "passive" phenomenology of time experience with Heidegger's "active" conception. I believe that the idea of "timing" could help us to address one important question that he finds bothersome yet, as far as I can see, leaves open: Should this whole problem of narration be considered merely as "literary"?[27] Not at all, I would argue. The most literary story still involves action on the part of its author. Action requires timing; telling a story above all involves choices between what is said, discursively and explicitly, and what is unsaid but nevertheless "stated" by the narrative structure. That narration is also a matter of timing is the deepest reason why "rhetoric" is linked to power and why, given the power relations that exist between narrators and audiences, "time to tell a story" is a contested resource and why "constructions with time" used to give accounts of relations between cultures need to be looked at critically.

La Fontaine and Wamimbi

The Anthropology of 'Time-Present' as the Substructure of Historical Oration

DAVID WILLIAM COHEN

IN HIS RECENT COLLECTION OF essays, *Time and the Other: How Anthropology Makes Its Object,* Johannes Fabian argues that "Time as a dimension of intercultural study (and praxis) was 'bracketed out' of anthropological discourse."[1] He sees the encapsulation of Time—as an object of study—as a creature of, and a problem for, anthropology. He argues that the relativistic or comparative approaches to the study of Time have led communities of anthropologists to reify typologies of Time while setting other cultures within distinctive, and *other,* conceptual frameworks of Time. Fabian sees these relativistic constructions of Time, which he takes to be a significant aspect of the conventional representation of other cultures, as rhetorically different from but essentially similar to the designation in other literatures of other cultures as "savage," as "primitive," and as "underdeveloped."

For Fabian, the ways in which anthropologists have proceeded to study Time, and to represent the Time of others, as *other,* are little recognized but nevertheless significant aspects of the organization of power or domination in the world. As a central thesis of the book, Fabian critiques anthropology for its failure to represent the fact that, at the critical moment in fieldwork, anthropologists and their subjects of study share one Time, not different temporalities. The study of Time in other cultures, in multiple cultures comparatively, is not attacked by Fabian as a continuation of an antiquarian collection of interesting ethnological notes that mark cultural difference. Rather, he attempts to show how anthropology, far from having control of its procedure, reflects more fundamental tendencies in Western thought toward what he calls "visualism,"[2] and in Western practice toward a will to power.

African historians have for a long time lamented the ways in which many anthropologists doing pioneering work in different areas of the continent eschewed the study of past, even as they might occasionally offer rich expositions on, and attach considerable significance to, "native systems of time-reckoning." Lamentations aside, the distinctive way in which anthropologists tended to approach African societies and cultures in the period between 1930 and 1965 permitted new schools of historians of the continent to adapt what they saw as "time-less" descriptions of society and culture to the challenge of reconstructing the past. The historians' results could be extremely sophisticated merges of research materials and methods with clearly different objectives. Or they could be extremely simplistic and problematic attachments of a diachronic dimension to a synchronic cultural and sociological description. Or they could be, more simply, just hostile reactions to the purposes and resulting descriptions and analyses of anthropologists.

The considerations of the African historian, like those of Fabian, have much to do with the discourses of guild scholars based outside of Africa: how a guild's models of study of Africa force a discourse within the guild and between guilds. For Fabian, as for James Clifford and others who are concerned to identify what they see as powerful tropes and conventions within ethnography as text,[3] the critique of recent and contemporary anthropology is voiced in terms of a campaign to empower the subjects of its craft, or as Fabian notes, its "object." These critics and textualists call for close examination of the paths that anthropologists take toward their topics, for new styles of presentation, for coauthored text, for self-critical reflection on the specific project of fieldwork, for empowering the voices of the subjects of study.

For all the claims to destooling the authority of ethnography and the ethnographic mode, these new critics move through their examinations of texts as if anthropology were some autonomous discourse working only upon itself.[4] These projects—authorship with subjects as coauthors, the empowering of informants' voices, the transposition of multiple voices from public discourses into text—are, in the end, orchestrated by the anthropologist. It is noteworthy that on the cover of James Clifford and George Marcus's collection of essays in this vein, *Writing Culture,* the European field-worker is shown in the foreground busy with his notebook and the presumed informant somewhat opaque, removed, and bored in the background.[5]

In his 1986 essay in the collective work *Writing Culture,* Jim Clifford remarks on what he sees as a recent and significant development, native authorship of alternative studies of societies and cultures:

> "Informants" increasingly read and write. They interpret prior versions of their culture, as well as those being written by ethnographic scholars. Work with texts—the process of inscription, rewriting, and so forth—is no longer (if it ever was) the exclusive domain of outside authorities. "Nonliterate" cultures are already textualized; there are few, if any, "virgin" lifeways to be violated and preserved by writing. . . . A very widespread, empowering distinction has been eroded: the division of the globe into literate and nonliterate peoples. This distinction is no longer widely accurate, as non-Western "tribal" peoples become increasingly literate. But furthermore, once one begins to doubt the ethnographer's monopoly on the power to inscribe, one begins to see the "writing" activities that have always been pursued by native collaborators.[6]

In fact, Clifford demonstrates little interest in the relationship between native authorship and the ethnographic text. Significantly, what he terms the "intertextual predicament" arises in his argument not from the contest between native and guild text but rather from the situation in which the field-worker faces a field full of prior written accounts, a "field . . . already filled with texts."[7]

The study of the relationship between the so-called "native text" and the "guild text" has hardly begun, perhaps because in the end the ways in which an "ethnographic text" may empower or disempower a "native text" are only of peripheral interest to the enclosed anthropological discourse over the authority of the guild text within anthropology. Conversely, the ways in which a "native text" may empower or disempower an "ethnographic text" would raise alarms in the guild concerning the conventions of authorship, originality, and the primacy of field observation, dislodging such a derivative text from the genre of "ethnography."[8]

What is lost in the disregard of the relationship between native text and guild text is an understanding of the potential power of the conventional ethnographic mode in unintended processes of the production of knowledge, an understanding that the most conventional of ethnographies may empower the subjects of ethnographic inquiry. To argue—as Clifford and his colleagues do—that new forms of ethnographic representation must be created that are to be designed to empower the subjects of such inquiry is to miss the paradoxical and often ambiguous and unpredictable nature of domination and response in the world we inhabit.[9] Whatever Clifford and his colleagues may intend by the language and mode of their discourse, they cannot close off the unintended channels through which individuals and groups draw out, construct, and command knowledge about themselves and others. As in the production of history, the production of ethnographic knowledge outruns the controlled discourses of the guild.[10]

As Johannes Fabian has elegantly argued, Time has an essential place in this discussion because it is by its consensual construction of Time around a premise of "anthropological time-present" that conventional ethnography has marked its own discourse. It is, as Fabian helps us to understand, in the conquest of Time that ethnography most characteristically and formidably constructs its texts. On the other hand the author of a native text may find her or his authority not only in a prior ethnographic text (no matter how conventional), but also in the reconquest and reconstitution of Time within the production of the native text itself.

The relationship between the "guild texts" of Jean La Fontaine, a Cambridge-trained anthropologist, and the "native texts" of G. W. Wamimbi, a Ugandan secondary-school teacher, is an excellent case in point.

In 1959, Dr. Jean La Fontaine, a British anthropologist, published *The Gisu of Uganda,*[11] a volume in the series Ethnographic Survey of Africa. This thin volume was based on anthropological fieldwork conducted between 1953 and 1955 among the people of eastern Uganda known as the Gisu; the author also produced a dissertation for Cambridge University (the Ph.D. was conferred in 1957; the dissertation was never published). *The Gisu of Uganda* and subsequent articles on the Gisu were, as a collection, an important contribution to British social anthropology, particularly in the study of initiation ritual. These studies reflected, at the same time, the formative influences upon La Fontaine of the field as it was constituted in the early 1950s, and particularly the interests of her mentor Meyer Fortes in comprehending the ways in which roles become, through ritual, transitional structures.[12]

Like the studies of other social anthropologists of her day, La Fontaine's work has been characterized as ahistorical or, as some African historians glibly yet perhaps aptly remarked in reaction to other ethnographies, "anti-historical," clearly demarcating an ethnographical vision from a historical one. During a seminar at The Johns Hopkins University in December 1985, La Fontaine challenged the value of oral evidence for historical reconstruction several times, with specific reflection on the study of the Gisu past, declaring oral traditions and testimonies "suspect" and "ingenious" and "ungrounded in reality." When asked about the past of the Bagisu before 1954, La Fontaine referred the questioners either to written sources produced by colonial officials and missionaries or remarked on the paucity of written sources, which "regrettably limited the opportunities for doing proper history."[13]

The assertions of La Fontaine concerning the sources and possibilities of Gisu history and her consistent, indeed expert, use of the "anthropological time-present" and an alternate "anthropological time-past" in her ethnographic writings are simultaneously distinct and interlocked formulae. Notably, in a later and more general treatment of initiation—*Initiation: Ritual Drama and Secret Knowledge Across the World*[14]—La Fontaine switches back and forth between an "anthropological time-present" and alternative temporal markings, even using specified years within her field research to denote specific observations. Almost as if he were observing La Fontaine begin a transition from the controlled use of the "time-present," James Clifford has observed that a generation (the span of 30 years between the close of La Fontaine's Gisu fieldwork and the date of publication of her general volume on initiation) is "approximately the temporal distance that many conventional ethnographies assume when they describe a passing reality, 'traditional' life, in the present tense."[15] The expression that we use here—"time-present"—does not mean present time. It is, rather, an artificed synchronic temporal form. As we shall see, to the eye of the reader of La Fontaine's early writing on the Gisu, she in fact utilizes two temporal forms: a "time-present" encompassing the meanings of "traditional" and "continuing" and a "time-past" carrying the meanings of "previous" and "earlier times."[16]

In April 1970, G. W. Wamimbi, a Mugisu schoolteacher, produced a long and detailed English-language manuscript entitled *Modern Mood in Masaaba.*[17] Wamimbi was one of the founders of the Masaaba Historical Research Association, and *Modern Mood in Masaaba* was produced as an eloquent oration of the history and culture of the people presented in La Fontaine's study as the Gisu or Bagisu. In Chapter 8 of *Modern Mood,* Wamimbi presents an extended, detailed discussion of male circumcision (*Imbalu*) among the people of Masaaba. For Wamimbi, as for La Fontaine and for others who have written on the Gisu, *Imbalu* is the central rite of the people of Masaaba. Where La Fontaine's *Gisu* presents four pages on *Imbalu* (pp. 41–45), Wamimbi's text is perhaps twice the length.

What one notes, in setting the La Fontaine and Wamimbi scripts side by side, is that the La Fontaine text of 1959 and the Wamimbi text of 1970 are clearly related. Wamimbi appears to have paraphrased La Fontaine rather liberally. The Wamimbi text reads as if it sits firmly upon the authority of the La Fontaine; yet in an interesting way Wamimbi has expanded boldly on the La Fontaine text at a number of points (thus the greater length). He has, without explaining his rhetorical operation, adapted the La Fontaine text to his own purposes. The La Fontaine text has empowered Wamimbi's own text, and one may argue that La Fontaine has actually made possible the Wamimbi text. This is not to exclude the possibility that La Fontaine and Wamimbi might have both been influenced separately and serially by still earlier treatments on *Imbalu* not cited by either.

Importantly, Wamimbi has, in his writings, begun the reconquest of Time. Wamimbi's reconquest of Time is both explicit and implicit. For one thing, he introduces considerable temporal specification into the La Fontaine treatment of *Imbalu*. But also, where he utilizes the present tense, it may mean both more and less than La Fontaine's "time-present." For another thing, Wamimbi has an unconscious and consequential disregard for the conventions in ethnography concerning the use of the "anthropological time-present"; in an important way he helps us see La Fontaine's "time-present" as artificed synchrony. Moreover, Wamimbi has a longer span of time to establish the currency or continuity of aspects of *Imbalu* as described a decade and a half earlier by La Fontaine. And Wamimbi establishes in his text very powerful concerns, indeed a political and cultural agenda, relating to the distinctions between present and past.

La Fontaine and Wamimbi: The Texts Compared

Our parallel and comparative reading of La Fontaine and Wamimbi begins in the second paragraph of La Fontaine's section on "Circumcision" in *The Gisu of Uganda* and in the seventh paragraph of Chapter 8 of Wamimbi's *Modern Mood in Masaaba,* but it should be noted that where Wamimbi devotes his first three paragraphs to the telling of a history of *Imbalu,* La Fontaine offers nothing of this kind. One will note from the first sentences below the way in which the 1970 Wamimbi text (to the right) liberally paraphrases the 1959 La Fontaine text (to the left).[18] Yet it is evident that Wamimbi expands markedly on, and from, the La Fontaine writing.

The rite of circumcision was performed at different times in different lineages, which had their traditional order for the performance of the ceremonies in the preceding lineage. The order was not rigidly fixed but, in general, was as follows: Bu-Ngokko, -Soba, -Siu, Sano, -Shiende, -Wagogo, -Tta, -Gobero, -Tiru . . . (La Fontaine, *Gisu,* p. 41)	The rite of circumcision is done at different times in various clans according to the wishes of the clan elders or basukhulu. In the days gone by circumcision was performed almost from clan to clan and in that way a lot of useful time was wasted. Really people in Bugisu cannot dare work if there is joy, dancing, singing and drumming in

the country side. People feel free and feel happy. When the Government realised this, they stepped in and each sub-county or Gombolola was given a day on which circumcision must take place. Sundays were excluded. The Bamutoto clan of Bungokho must always of necessity open up the circumcision period because this important custom began in their clan, and so the whole of Central Bugisu comes first before the rest of the District. Circumcision is taken in the following order on the days and dates actually specified by the Bugisu District Administration: *Central Bugisu:* Bungokho, Busoba, Busiu, Bukhiende, Bufumbo, Nakaloke . . . (Wamimbi, *Modern Mood,* Chap. 8, para. 7, pp. 2–3)

The rituals take place after the main harvest, which lasts from June to September, but formerly if there were a bad harvest the ceremonies would have been postponed a year. Today they are held regularly every other year, but it seems likely that formerly they were held less often. (La Fontaine, *Gisu,* p. 42)

In all cases circumcision always takes place after the main harvest of millet (bulo) particularly between June and September. But when there is very serious shortage of food ceremonies can be postponed until further notice. For instance there was very serious famine between 1917 and 1919 and because of the seriousness of that famine (*inzala iya Kakutiya*) boys and the uncircumcised men were not circumcised in 1919 which was supposed to be the year of circumcision, and so changed the order from odd years as it was formerly done to even years up to the present moment. (Wamimbi, *Modern Mood,* Chap. 8, para. 7, p. 3)

From this first extract into the second, La Fontaine shifts directly from a past tense into the "time-present," and in both instances she holds to a most general temporality, as if the precise specification of the boundary—the historical contingencies—between an "anthropological time-present" and an "anthropological time-past" were of no special significance. Wamimbi, for his part, begins with a "time-present" in the first sentence of the first extract, reversing La Fontaine's tenses, then boldly recovers Time in the second extract. By transforming La Fontaine's general temporality into the specific reporting of government edicts, and by introducing specific dates, he alters the nature of Time in the La Fontaine passage, making it possible for him to assert the significance of historical contingencies.

What is also evident at the outset of the first pair of texts is that where La Fontaine asserts the power of the "traditional" in establishing the order of timing of rites from lineage to lineage, Wamimbi revises La Fontaine and introduces "agency": "the wishes of the clan elders." The Western sense that Africans governed their lives by "tradition" and "custom" is confounded as we notice that it is the Cambridge-trained anthropologist who makes the case for the determination of the "traditional" while the local writer revises her work to introduce an entirely different perspective. And Wamimbi goes further, alerting the reader to the opening of a contest over the value of Time in the colonial period with the confrontation between the effects on the use of time rendered by the agency of clan elders and the notion that "useful time was wasted," which brought the "Government" to introduce regulations over the time of the rites. What is perhaps intended yet implicit in Wamimbi is an acknowledgment of an economy of circumcision, which the new "Government" policy affected: circumcision periods were made more uniform; the peripatetic character of the circumcisor's activity was substantially reduced; a shortage of circumcisors became evident; and a certain anxiety was noted in regard to the speed of cutting required for the circumcisors to get the job done.[19]

In the third extract, La Fontaine shifts back to a past tense, but her text does not reveal the grounds for this switch. Interestingly, what occupies some past time for La Fontaine in 1959 is in the present in the 1970 text of Wamimbi. One might suspect that for La Fontaine these sentences describe activity that she viewed as secondary to events on the central stage of circumcision ritual. The past tense perhaps also reflects a sense on her part that a once-important practice—the exchanging of gifts between age-set mates—was being peripheralized as a result of African Local Government suppression of the practice. Wamimbi's present tense, however, denotes a possibly significant political valence attached to the gift practice. Where La Fontaine's treatment is an epitaph reporting the disappearance of a cultural practice, Wamimbi's text begins and ends as an oration on its high and continuing importance—though in the discussion of rules established within, as opposed to those pressed into service by Government, he shifts to the past tense.

Set next to the valorizing and prescriptive sentences of Wamimbi, La Fontaine's text suggests the chilling violence of description in the "timepast." Meaningful practices are interred by this tense. On the other hand, Wamimbi struggles to revive the past and hold back the present. His reconstitution of tense forms allows him to introduce and assert affect; this is substantially absent from the La Fontaine passages. But, importantly, Wamimbi seems to be after more than simply a romance with the past. Here and elsewhere, Wamimbi uses tense demarcation as a means of distinguishing the realm of Government program and influence from internally sustained elements in the ritual.

Men circumcised in the same year, and particularly in the same village, were expected to behave towards each other as brothers. They called each other by a special term, *magoji,* and might not marry each other's daughters. The children of two *magoji* might not marry, but the grandchildren were allowed to do so. *Magoji* were expected to make certain gifts to one another, of which the most important was a cow at the circumcision of the son of one of them when he has the obligation to feast his age-mates with meat and beer. The ruling of the African Local Government in 1954 that there were to be no more such gifts caused widespread complaint. (La Fontaine; *Gisu,* p. 42)

The boys and the men who undergo this intolerable and unforgettable pain in the same year regard themselves as real brothers—Bamakoki or Magoji. We can regard them as people of the same age-group. They are not allowed to marry each other's daughters and even their children may not marry one another. This tradition is very strong and began at the time when circumcision began in Bugisu. There are other rules and beliefs in connection with Bamakoki. They are not allowed to wrestle with one another. It was forbidden for these men to meet in the latrine or any other places of excretion. It was unlawful for them to find one another on the roof of a hut thatching the roof. It was considered illegal for one to find his Magoji bathing at the stream. If one of the above rules was violated, then the offender was asked to pay a fine to his Magoji. The fine was to be negotiated between those concerned. Usually amicable agreement was reached. A man was almost compelled to give his Bamagoji things in the form of money, meat, chickens and beer if he had a son who had been circumcised that year. In some parts of the District, this practice is still carried on a very high scale. It is therefore important to remember that circumcision as a custom of the Bamasaaba creats [*sic*] a sense of fellow-feeling among men who are circumcised and initiated during the same year and these same people sincerely retain a sense of corporate identity. (Wamimbi, *Modern Mood,* Chap. 8, para. 7, p. 4)

In the fourth extract, La Fontaine turns to John Roscoe's 1915 work *The Northern Bantu* for assistance in developing a description of the circumcision ritual. "The account here," she writes, "is based on personal observation, with references to Roscoe where significant variations occur" (*Gisu,* p. 42). There is some irony in the fact that as Wamimbi (who himself went through the circumcision ritual and had occasion to participate in and observe perhaps a dozen cycles up to 1970) turns to La Fontaine (who observed but one cycle during her two years of fieldwork) for description, so La Fontaine turns to John Roscoe, who, 40 years earlier, though a fine observer, was no expert on the Gisu specifically. It is upon noting this sequence of dependence in the series of observations of *Imbalu* that one commences to ask if it would have been possible for Wamimbi to write his "native text" without having the "guild text" on which to build his own and from which to draw the authority that his political agenda required.

The first phase is occupied entirely with dancing. The novices dress up in the traditional costume and dance on the village greens, under the direction of an older, already circumcised man, who leads the songs and instructs the novices in the dances and songs they will have to sing. Roscoe states that the initiates are given formal instruction by the el-

The intending boys traditionally dress and dance in their local villages at special places. They wear bells—bizenze or zikuuma—which are always worn on the thighs or kamakimba which are always used in the hands. The majority of the boys in Bugisu use the former type of bells. But kamakimba are used chiefly in Bumbo and Bubutu, but

ders before circumcision, but I did not observe this and I do not think it ever occurred. The dress of a novice consisted of a headdress made of the skin of a colobus monkey; cross-bands made of the nuts of a tree or beads used by women for the girdles worn by all women round the hips under the main clothing. (La Fontaine, *Gisu,* p. 42)

they may be used in Butiru or Buwabwala. The first type of dance is called "*isonja.*" The boys hire an expert man who is already circumcised—usually called by different names of Khyilali, Namwenya or simply Umwimbi according to various parts in the District. Besides bells worn by the initiates, they wear a head-dress—lilubisi—made of the skin of a colobus monkey; then cross-bands from special seeds or nuts of a particular tree—kamaliisi or beads formerly known as madongo. (Wamimbi, *Modern Mood,* Chap. 8, para. 7, p. 4)

La Fontaine places her descriptions of dance in the present tense in extract four, but she goes to Roscoe and the past tense for the description of dress. Wamimbi reintegrates the dance and the dress of the dancers through the use of the present tense for both.

In the fifth pair of extracts, where La Fontaine simply demarcates a present time and a time "formerly," Wamimbi employs tenses that allow him to remark on transitional processes in the transformation of both the dress and the dance. Where La Fontaine offers no evaluation of this distinction between a present pattern and a former one, Wamimbi laments the transition in practices and what he sees as losses under way. And, lastly here, where La Fontaine talks about the dancing activity in a general temporal sense (no special time), Wamimbi locates this as "usually [taking] place between 3 P.M. and 7 P.M."

These strings of beads are given to the initiate by his "mothers" and "sisters" to give him strength; a belt of hide sewn with cowrie shells and occasionally decorated with tassels of monkey-skin as well; in the south, two strips of hide, sewn with cowrie shells, are worn hanging down the back; iron thigh-bells are worn on one or both legs, according to local preferences; iron bracelets are worn on each arm and a circlet

Some of the boys used belts of hides sewn with cowrie shells (zisimbi) accompanied by other decorations according to the wishes of the initiate. In some parts of the District, some boys wear two strips of hide beautifully sewn with cowrie shells which hang down the back. They are called "kamatondo." The thigh-bells can either be worn on both legs or on one leg. Iron bracelets are worn on each arm and two circulets of wood higher up on the upper arm. A hippopotamus tusk was tied to the headdress over the forehead. These tusks were, and are, family heirlooms handed down carefully from generation to generation. Now they are very rare and are sometimes imitated in wood. Formerly the novices were naked, except for these ornaments, but today they wear shorts and cheap ornaments—bright bangles, handkerchiefs, etc., bought from shops. (La Fontaine, *Gisu,* p. 42)

[*sic*] of wood are worn high up on the upper arms. These are known as "sipogo" or ( ). Some boys especially big boys do wear a hippopotamus tusk, called igweena which is tied to the head-dress over the forehead. These particular tusks are carefully kept and then handed down from generation to generation in various clans. It is unfortunate that these traditional zikweena are gradually diminishing as the time is changing. In the days of old and before the coming of the white man to Uganda, boys wore no shorts save the ornaments as described here; but today every initiate must wear shorts and some or all of the said ornaments plus bright bangles, handkerchiefs and other things of decorations. It is further regretted that the isonja dance which usually takes place between 3:00 P.M. and 7:00 P.M. is being neglected in parts of Bugisu for various reasons. But indeed it is a very good, fascinating and preparatory dance for the intending initiates. It is the time when boys acquaint their thighs with bells. The isonja dance usually takes place between March or April and June. (Wamimbi, *Modern Mood,* Chap. 8, para. 7, p. 5)

There are a further fourteen pairs of related sections of text dealing more closely with the operation and with postoperative rituals, ceremonies, and practices. Lack of space precludes our considering them all here; so we will skip to a closing pair of extracts to see how each author deals with the operation itself.

The circumcisor or circumcisors then come forward and, beginning with the most senior, circumcise down the line as fast as they can.

The circumcisor called Umukhebi instantly steps forward to begin the physical operation. If there are many boys at one particular yard,

The novices must stand absolutely still and not show, even by the flickering of their eyelids, that they suffer. The spectators, both men and women, keep up a deafening noise of shouts and exhortations. When the foreskin has been cut off there is a shout, "They have spoiled you!" (La Fontaine, *Gisu,* p. 44)

the circumcisor begins with the most senior. The Umukhebi is helped by his assistant called Umuhambi or Umubinjilili. He circumcises down the line as fast and yet as carefully as he can. The boys are, of course, expected to stand absolutely still looking in one direction in front of him. They must not show or even blink their eyes. At this moment the onlookers keep up shouting and yelling and the atmosphere is mixed with great excitement full of joy, worry and great expectations. The noise made by the spectators is extremely deafening. Immediately, the foreskin is cut off people shout at us [*sic*], "They have spoiled you!"—"O! Bakhwonaka." (Wamimbi, *Modern Mood,* Chap. 8, para. 7, p. 8)

The Negotiation of Time

La Fontaine, at the Hopkins seminar, saw her source materials as without value for the reconstruction of the past, and at no point made any claim that her ethnographical writings could be a basis for historical exposition. Yet Wamimbi saw her materials, as digested into an ethnographical genre, as an important means of valorizing the past. Where La Fontaine appears to have considered those stories that were presented to her as speaking to present social and cultural concerns rather than as opening toward a reconstruction of the past, Wamimbi saw La Fontaine's materials as making possible a strong and controlled appreciation and powerful use of history among the Bagisu. Those African, American, and European scholars at the December 1985 *Imbalu* workshop at Johns Hopkins heard La Fontaine's perspective on her own texts as a distinctly and enduring ahistorical one, while Wamimbi saw La Fontaine's *Gisu* as a powerful basis for producing history.

Wamimbi made sense of La Fontaine's analysis not through an appreciation of "structures" and "functions," but through a realization of the openness of Bugisu ritual. Wamimbi produced in one sense a "double narrative." He worked within the conventions of presentation of La

Fontaine's ethnography, building upon the authority of her printed work; yet he sought to produce another and very different work, with a concern for presentation of the constraints and contingencies that produced variation and made vulnerable what he considered valued elements of circumcision ritual. He was also concerned to reintroduce into the treatment of circumcision the emotional, or affective, aspects of the ritual, the tumultuous qualities and pleasures and pains of experiencing circumcision as audience and as initiate.[20]

La Fontaine's systematic treatment of Time, with her exclusion of the historical contingencies lying between an "anthropological time-present" and an "anthropological time-past," may have engendered a stronger sense of the power of Time and of historical contingency in Wamimbi's work, raising his consciousness of the significance of Time in the presentation of the circumcision ritual and perhaps occasioning the construction of what is read here as a historical oration. Although La Fontaine reworked Time in accord with the conventions of the ethnographic report of the day, Wamimbi was unable to sustain this structuring principle of her text, for he was too aware of the continual contestation of issues of *Time*.

Some of this contestation is related in the examples of Wamimbi's prose offered above, but there is another salient field in which Time is at issue: the extended record of named age-sets. Most of these lists extend back to the early nineteenth century and thus record a sequence of circumcision episodes from what is held to be the first Gisu circumcisions up to the present. These lists are reported by a number of observers of the Bagisu. In her doctoral thesis, La Fontaine reports one such list, and that report was sufficient for her purposes. But Wamimbi saw this as an important area of interpretation and contention. Wamimbi perhaps saw that each list could be read properly only by reference to the information contained in other lists. The lists vary in their detail not only from text to text but also according to which area of the Gisu countryside the particular list, or list fragment, refers. In *Modern Mood* Wamimbi produces one of the most elaborately annotated list-texts available from Bugisu.[21]

In *The Chronology of Oral Tradition,* the historian and archivist David Henige examined ways in which "oral traditions arose in response to a broad range of stimuli."[22] Henige took the printed word as one of these stimuli, a very significant one, and adopted the term "feedback" to refer to the process by which oral tradition was influenced or remade by the written and published word. Epigrammatically, Henige cited a Chinese proverb: "The most retentive memory is weaker than the palest ink." Henige himself wrote: "Feedback may be defined as the co-opting of extraneous printed or written information into previously oral accounts. The process occurred very widely, if not obviously, and its prevalence emphasizes the extent to which newly literate societies recognized the truth of the Chinese proverb."[23]

Henige provided numerous examples of accounts from around the world that suggested to him the power of the written and published word for people not yet, or just recently, exposed to literacy. Henige's examples cite and incorporate written sources, often newly written and sometimes inexpert ones. Henige notes: "Examples of this propensity to incorporate the written word into putatively oral traditions, or to base these 'traditional' accounts entirely on printed or written sources can be found wherever literacy has occurred in Africa, not only in areas of European language penetration but of Arabic influence as well."[24]

Henige's point in this discussion, and in a later work, *Oral Historiography,*[25] is to demonstrate the vulnerability to the written word of what he has taken to be pristine oral tradition. Henige warns that the historian using oral tradition can fall victim to a circular effect in the production of the historical source and that the facts the historian seeks to uncover from oral sources may actually derive from less expert, possibly interested sources from outside the society: missionaries, explorers, educators, government authorities, and so forth. At a deeper level, ideas and structures may be lifted from books, most particularly the Bible, and may then set the structure and content of oral tradition or of literary works based on oral tradition.

We may wish to suspend our concern for the questions arising from the presumptuous creation of a dichotomy between written sources and oral tradition. Historians and others interested in contact between literary and nonliterary traditions can hardly live with the simplifications and reifications involved in the Henige dichotomy, whatever the recognition of influence of one source upon another might be. Henige took us into the subject, but not very far into the discourses that develop in innumerable settings between source and source, between book and memory, between a written record and an oral testimony, between a work of guild scholarship and a reformulation of that work for a different audience.[26]

A useful argument, and one in sharp conflict with Henige's view of the influence of the written word on the oral, may be drawn from Carlo Ginzburg, *The Cheese and the Worms: The Cosmos of a Sixteenth Century Miller.* Ginzburg's subject, Menocchio, is discerned to be an avid and eclectic reader. Ginzburg explores the relationships among the ideas about the world that Menocchio articulated (and reproduced for the Inquisition) and published sources and knowledge circulating through a still existing oral culture:

> In Menocchio's talk we see emerging, as if out of a crevice in the earth, a deep-rooted cultural stratum so unusual as to appear almost incomprehensible. This case . . . involves not only a reaction filtered through the written page, but also an irreducible residue of oral culture. The Reformation and the diffusion of printing had been necessary to permit this *different* culture to come out to light. Because of the [Reformation], a simple miller had dared to think of *speaking out,* of voicing his own opinions about the Church and the world. Thanks to the [diffusion of printing], *words* were at his disposal to express the obscure, inarticulate vision of the world that fermented within him. In the sentences or snatches of sentences wrung out of the books he found the instruments to formulate and defend his ideas over the years.[27]

In rather less stunning ways, there are resonances with Wamimbi's production of a unique, and historically sensitive, oration out of the words and structure of a carefully fashioned and ahistorical ethnographic text.

In a 1982 essay, Roger Chartier brings attention to Ginzburg's close study of Menocchio as a reader of diverse sources and to the significance of *reading* in the passage of knowledge from one field or system to another:

> As Carlo Ginzburg has shown, what readers make of their readings in an intellectual sense is a decisive question that cannot be answered either by thematic analysis of printed production, or by analysis of the social diffusion of different categories or works. Indeed, the ways in which an individual or a group appropriates an intellectual theme or a cultural form are more important than the statistical distribution of that theme or form.[28]

The close study of Wamimbi's reconstitution of La Fontaine is one path of entry into Wamimbi the reader. A brief look at Wamimbi's career offers another.[29]

It is clear from what we know of Wamimbi's life that the discourse between the La Fontaine text and the Wamimbi text was not a simple one of adaptation of ethnographic data from one source into another. Wamimbi was at the helm of the premier historical research group in Uganda in the 1950s, 1960s, and 1970s—the Masaaba Historical Research Association. This was an improvement society for young educated men. It was founded as a school society by a group of boys in their mid-teens.

In an interview conducted by Stephen Bunker on March 7–8, 1970, Wamimbi said, according to Bunker's interview notes, that "he had started MHRA in 1954 with a group of other students who were in J3 with him at Nyondo TTC [Teacher Training College]. He was then 17. In 1956, the MHRA was extended from a student organization to include adults, but Wamimbi remained President, and has been since."[30] On January 10, 1962, the association drafted and approved a constitution, with these stated "objects": "to collect the past and present history of Bugisu and to establish Masaba Museum; to write books in Lumasaba; to collect and correct common mistakes and develop Lumasaba grammatically; to translate various books from other languages into Lumasaba; to find meanings of names of places and people of Bugisu; and to write down the Geography of Bugisu."[31] Within ten months, the Association was petitioning the Minister of Regional Administration in Entebbe, in a memorandum entitled "Bugisu Land Belongs to the Bagisu." Following a preface on the geography and the early colonial history of the region, the association wrote:

> The Bugisu-Bukedi Boundary Dispute Commission was set up early this year. We willingly co-operated with the Commission but what has happened to its report? The Bagisu are thirsting and longing for the Report. There is a rumour that Bukedi, Bugisu and Sebei Districts might be re-amalgamated. Indeed that is very unacceptable and can never be entertained by any Mugisu. We shall bitterly and energetically oppose that move.

In December 1965, Wamimbi participated in a seminar on "the techniques of collecting and recording local history" at Makerere University College. In his presentation to the meeting he reviewed the objectives of the association, among which were to preserve the "Past and Present history of the Basmasaaba and to establish a Museum for all or some of the Kimasaaba materials used by the Bamasaaba. . . . In this first aim we devide [*sic*] our history into (1) That before the Europeans set foot in Bugisu and (2) that from the coming of Kakungulu and the Europeans in Bugisu and so on." Wamimbi went on to specify the linguistic agenda, expressing concern to establish a standard Lumasaaba orthography as a preface to writing and publishing books in Lumasaaba. He then proceeded to the subject of circumcision:

> Although we have known some facts on *circumcision,* a lot has to be done, for its ceremonies differs [*sic*] almost from sub-county to sub-county. Circumcision is by far the most favourable custom in the District. It can be compared to the Olimpic [*sic*] Games of Old Greece. Every Mugisu young or old,

> female or male must like it and every person is . . . forced to join it in the dancing and running. (It is a wonderful Scene when the time for circumcision has come or is approaching.) There is peace and joy all over the District during this time of the year. People are happy, impatiently awaiting the occasion.[32]

Wamimbi then reviewed at length further projects, achievements, and methods, and brought attention to specific plans to cooperate with parallel organizations elsewhere in East Africa.

On December 30, 1968, Wamimbi wrote a long memorandum to the Bugisu District Administration, which had said that more information was required before fresh funds could be provided to the Masaaba Historical Research Association. Again, Wamimbi reviewed the objectives and achievements of the association, this time referring to regional conferences held in previous years in the district. More interesting for the present work was the inclusion of a brief report on the progress of a book being produced by the association on the "History of Bugisu," to be titled *Bugisu in Modern Uganda*. Appropriately, Wamimbi offered the District Administration no account of the political agenda of the association. But there was, indeed, a considerable political agenda: recognition of the Lumasaaba language and standard orthography; the reformation of the school curriculum in the Masaaba area; the reparation of boundaries in eastern Uganda; the organization of district administration; the adjudication of disputes with the government over the timing, regulation, and legality of elements of the circumcision campaigns. For Wamimbi and his colleagues in the association, all these problems were addressable through arguments from history. In the years following, the association produced a 72-page *History,* a series of special papers entitled "Malembe" or "Masaba Historical Research Association Presents Its Work," and Wamimbi's *Modern Mood in Masaaba,* with the treatment of circumcision based on La Fontaine's account.

A question arises as to whom this work was addressed. Who constituted the audience for Wamimbi's text? A first point is that the Wamimbi text was drafted and reproduced in English. Here Wamimbi sat amidst a tangle of contradictions inherent in colonial and postcolonial intellectual life. To address questions of power or authority in the defense of culture, territory, local governance, ethnicity, one had to speak or write in the language of the colonial power or, with independence, the national government. To reach authority Wamimbi could not at the same time address his own communities in their language. Wamimbi's position was the more complicated because to attempt to write in Lumasaaba or Lugisu before achieving an accord on orthography was to give preeminence to one dialect cluster of Lumasaaba and thus one subregion of Bugisu. Moreover, Wamimbi's prose may seem to be produced for, and directed toward, an audience of one: the magistrate or district commissioner. If Wamimbi "heard" La Fontaine, who indeed "heard" Wamimbi?

Given the specific contradictions that surrounded and marked Wamimbi's intellectual production, given the constraints on his writing in Lumasaaba, given his reading of La Fontaine, one may well ask what is the meaning of "native author" in this setting?[33] The dichotomy between "native" and "guild" broached earlier in the paper is open to challenge. The "native" character and authority that attaches to Wamimbi's text is not a consequence of his birth or of his socialization to a cluster of core values. To address with power the officials of the Ugandan government, Wamimbi seizes the authority that lies in the "native" voice. The "native" voice is here not given but constructed. In his December 1968 memorandum addressed to the district administration, Wamimbi wrote:

> I assume you are quite aware that however able and sympathetic the foreign observer may be, he can never fully interpret the thoughts and aspirations of another tribe or country. Therefore, the Masaaba Historical Research Association is the very capable organisation to write a thorough history of Bugisu as the majority of its members are Bamasaaba themselves.

In this sense, the Masaaba Historical Research Association was a complex and well-organized device to induce an authoritative "native" voice from the murky social and cultural field that was the Uganda of the late 1960s. What we see is that to produce his text, Wamimbi, the "native author," must command his own reading of an ethnographic text, reaching for its authority, but must simultaneously command the forms and values of "native" discourse within Uganda. To achieve his object, he comes to assemble these two textual modes, to make out of them a new and singular text with the doubled authority of the "native account" and the expert ethnography. The question of intertextuality moves to a more challenging plane. These are not simply two texts, one read and written out of the other, but a confrontation of modalities, of the ways knowledge is rendered, marked with authority, and transformed into literature.

The observed contention between Wamimbi and La Fontaine over the locations and values of Time opens to view the possibility that *Imbalu,* circumcision ritual, may be in a very important sense an extraordinarily rich and complex social and cultural negotiation over Time. Such a characterization offers a different and possibly revealing perspective on initiation. It may be compared to other approaches to the analysis of *rites de passage,* particularly approaches that stress the passage of the individual, individual psyche, and liminality. This latter cluster of approaches to the rite builds from the individual experience of pain as transition and as reformulation of the person.[34]

An approach that recognizes the significance of negotiation over Time would by nature emphasize the social and contractual character of the broad practice of initiation and the ways in which the young are remade through broader social practice. From this perspective, circumcision may be viewed as very "temporalistic." *Imbalu* was constructed out of the contentions over the precise timing that forms a structure for the ritual; the sense of time or of timing that each individual male has as he approaches the ritual; the elaborate discourse in larger Gisu society about the commencement of new circumcision ceremonies and the opening of new age-sets; the negotiation of holdings of knowledge of the time-series lists of age-sets by which contemporary Gisu locate themselves in relation to specific and carefully denoted moments in the past; and the coordination, adjudication, and contesting of the subregional timing of collective ceremonies. One might, indeed, see these elaborate interior discourses within Gisu society concerning the temporality of practices relating to the ritual as providing a collective ground for the transformation of the young.

This paper considers the location of time between two modes of reporting: one ethnography, the other history or historical oration. It examines the relationship between these two modes in one very specific and distinctive case in which the relationship between the two forms of reportage—the adaptation of one mode into another—reveals the distinctions between the two. But it is about more than a relationship between two modes of reporting. Some twelve years ago I was adjoined as a citizen member to the Johns Hopkins biohazards and radiation safety committee, which had just been charged with oversight of gene-splicing research in the Johns Hopkins institutions—not only to protect workers and the public from accidents but also to manage the confidential discussion of such research. Dangers lay not only in the biological materials that staff were adapting and creating but also in the fear and panic that would ensue were the enclosed and expert scientific discourse to escape from the lab into the public arena. Similarly, one might ask here what happens to the enclosed discourse on time in anthropology when the very stuff being managed "escapes from the lab"? The breach in the wall of control—the expectation that the scientific texts of anthropology would remain the stuff of anthropology—allows one to see the expert discourse on time and culture in a unique way as it is remade into popular voices and popular texts: the ways in which the scientific construction of "time-past" and "time-present" in ethnographic exposition are reconstituted in the practices of those observed.

Notes

Notes

BENDER AND WELLBERY: *Introduction*

1. Reinhart Koselleck, *Futures Past: On the Semantics of Historical Time,* trans. Keith Tribe (Cambridge, Mass., 1985).

2. Stephen Jay Gould, *Time's Arrow, Time's Cycle: Myth and Metaphor in the Discovery of Geological Time* (Cambridge, Mass., 1987); Ilya Prigogine, *Introduction to Thermodynamics of Irreversible Processes,* 2d ed. (New York, 1961).

3. Mikhail Bakhtin, *The Dialogic Imagination,* ed. and trans. Caryl Emerson and Michael Holquist (Austin, Tex., 1981), p. 84. For historical background, see Katerina Clark and Michael Holquist, *Mikhail Bakhtin* (Cambridge, Mass., 1984), p. 278; Gary Saul Morson and Caryl Emerson, eds., *Rethinking Bakhtin* (Evanston, Ill., 1989), pp. 23–24; and Gary Saul Morson and Caryl Emerson, *Mikhail Bakhtin: Creation of a Prosaics* (Stanford, Calif., 1990), pp. 366–69.

4. See Jean-François Lyotard, *The Postmodern Condition: A Report on Knowledge,* trans. Geoff Bennington and Brian Massumi (Minneapolis, 1984); and Niklas Luhmann, *Social Systems,* trans. John Bednarz (Stanford, Calif., forthcoming).

5. Donald J. Wilcox, *The Measure of Times Past: Pre-Newtonian Chronologies and the Rhetoric of Relative Time* (Chicago, 1987).

VAN FRAASSEN: *Time in Physical and Narrative Structure*

I wish to thank Professor Luca Cavalli-Sforza for help with the diagram; also Sally Haslanger, Margot Livesey, Gideon Rosen, and George Wilson for helpful comments.

1. Dino Buzzati, "The Seven Messengers," in his *Restless Nights* (San Francisco, 1983).

2. I speak here of the structure of the depicted events. There is of course in this, as in all stories, the rupture between the time structure of narration (*Erzählzeit*) and that of the narrated events (*erzählte Zeit*).

3. Every structural constraint can be violated, so some stories do not have narrated events all set in a single time structure. This is illustrated by John Fowles's *The French Lieutenant's Woman,* and its possible counterpart in reality is contem-

plated in Borges's "The Garden of Forking Paths." Margot Livesey, of Carnegie Mellon University, pointed out this limitation of the present paper.

4. *Sir Isaac Newton's Mathematical Principles of Natural Philosophy and His System of the World,* ed. F. Cajori (Berkeley, Calif., 1960), pp. 6, 8.

5. G. W. Leibniz, "Metaphysical Foundations of Mathematics," in *Leibniz: Selections,* ed. P. P. Wiener (New York, 1951); see esp. pp. 201–2.

6. B. Tomaschevski, "Thematics," in *Russian Formalist Criticism,* ed. L. T. Lemon and M. J. Reis (Lincoln, Neb., 1965), trans. p. 66; this passage is discussed by T. Todorov in *Introduction to Poetics* (Minneapolis, Minn., 1981), pp. 41–52.

7. Parfit once presented an example in which the declaration "But I am still in love with the same person, namely your 17-year-old self" is true; perhaps it would remain so if the speaker added, "and this would make me unfaithful if I loved your present self, for the two are not the same"—or something along those lines anyway. Cf. D. Parfit, *Reasons and Persons* (Oxford, 1984), p. 328, "She loves, not her middle-aged husband, but the young man she married."

8. Gérard Genette, *Narrative Discourse: An Essay in Method,* trans. Jane E. Lewin (Ithaca, N.Y., 1980), p. 43; the second part of the quotation is from Genette's n. 13.

9. By modern I mean roughly the period between the Renaissance and our century; it may or may not be objectively settled that we have that view now. In any case, my phrase "modern physics" does not cover contemporary physics.

10. Cf. the opening paragraph of Richard Feynman's *The Character of Physical Law* (Cambridge, Mass., 1967): "a rhythm and a pattern between the phenomena of nature . . . it is these rhythms and patterns which we call Physical Laws."

11. R. Barthes, "An Introduction to the Structural Analysis of Narrative," *New Literary History,* 6 (1975): 248; citation discussed by Todorov in *Introduction to Poetics,* p. 42.

12. See further Bastiaan C. van Fraassen, *An Introduction to the Philosophy of Time and Space,* 2nd ed. (New York, 1985), esp. chap. 3, sec. 3, and chap. 6, sec. 6.

13. The so-called quantum-logical interpretation keeps determinacy in this sense, but at the price of rejecting classical logic, a rather Pickwickian way of rescuing tradition.

14. I believe this to be so for all literature. However, as my colleague Sally Haslanger pointed out to me, it could be contested that this is so for reality as well. For example, Leibniz believed that every detail of reality follows logically from the mere premise that something exists. For he took it that God is limited only by logical impossibility, and also that He would not have created anything if there had not been a uniquely best possible world to create—and all of this he took to be logically necessary, since it was inherent in the concept of God. Of course, Leibniz may have been mistaken about this. I would point out that there are indeed familiar examples of finite axiom systems which are complete in the requisite sense; but this is always so within the confines of their own vocabulary, and they are non-uniquely extendable within a larger vocabulary. If that applies generally, then the world described in any conceivable story may be indefinite in ways not even hinted at in the story. But what if the story denied this of itself?

15. Genette, *Narrative Discourse,* p. 53 and n. 30.

16. B. Russell, *Our Knowledge of the External World* (New York: Norton, 1929), pp. 123–28. See also his "On Order in Time," *Proceedings of the Cambridge Philosophical Society,* 32 (1936): 216–28.

17. G. W. Leibniz, *New Essays,* 2.14.24, 26. in *Selections.*

18. Cf. Leibniz, *Selections,* pp. 231, 247, 253, 272–73.

19. Cf. Aristotle, *Poetics,* trans. R. Janko, with an introduction (Indianapolis: Hackett, 1987), pp. xxii–xxiii.

CASTORIADIS: *Time and Creation*

1. To facilitate reading, I have eliminated footnotes; in some cases, references are included in the main text. Here, I limit myself to some indications which may be of help to the interested reader.

I have developed the notions of the ensemblistic-identitary (which now, for brevity, I write ensidic) and of the imaginary mainly in my book *The Imaginary Institution of Society* (1964–65, 1975) now available in English (Oxford, and Cambridge, Mass., 1987). Especially relevant for the discussion of the present text are the Chapters IV (philosophical and social institution of time, identitary time as opposed to imaginary time, the social-historical as creation of a proper temporality), V (on the social institution of ensemblistic-identitary logic), and VII (on the social imaginary significations). To the ensidic logic I oppose what I call a logic of magmas; the idea was first formulated in "Modern Science and Philosophical Interrogation" (1971–73), included in *Crossroads in the Labyrinth* (1978), now available in English (Brighton, U.K., and Cambridge, Mass., 1984), in particular pp. 207–20. It was further developed in *The Imaginary Institution,* op. cit. pp. 340–44, and, in a much more detailed fashion, in "La logique des magmas et la question de l'autonomie" (1981) included in *Domaines de l'homme—Les carrefours du labryinthe II,* Paris, 1986, pp. 385–418.

On the ultimate inseparability of subjective and objective, see my text "The Imaginary: Creation in the Social-Historical Domain," in *Disorder and Order,* Proceedings of the Stanford International Symposium, September 14–16, 1981, Stanford Literature Studies Volume 1, 1984, Saratoga, Calif. [reprinted in *Philosophy, Politics, Autonomy* (Oxford, forthcoming)], and, in more detail, "La portée ontologique de l'histoire de la science," in *Domaines de l'homme,* op. cit., pp. 419–55. On self-reflective subjectivity see the First Part ("Psyche") of *Crossroads,* op. cit., Chapter VI of *The Imaginary Institution,* op. cit., and my text "L'état du sujet aujourd'hui," in *Topique,* No. 38, Paris, L'Epi, 1986 [now available in English as "The State of the Subject Today," *Thesis Eleven,* 24 (1989), pp. 5–43].

On the socialization of the psyche and the social fabrication of the individual, see *The Imaginary Institution,* op. cit., Chapter VI.

Paul Ricoeur's important book *Temps et récit,* three volumes (Paris 1983, 1984, 1985), is now available in English as *Time and Narrative* from University of Chicago Press; Vol. 3 was published in 1988. My obvious and central differences with Paul Ricoeur do not of course stand in the way of my admiration for the richness and solidity of his critical analysis of the main inherited philosophical conceptions regarding time.

On Aristotle's theory of time, the late Victor Goldschmidt's *Temps physique et temps tragique chez Aristote* (Paris, 1982) is fundamental. I do not always share his outlook, especially insofar as he tries to interpret away all riddles in Aristotle's text.

The recent book by the great physicist Stephen Hawking, *A Brief History of Time,* is sadly disappointing, and I am not surprised that it has been, for more than 100 weeks now, on the American best-seller list. It juxtaposes to an elementary and flat summing up of the history of the question of time in physics a wild final speculation, totally devoid of rigor.

One must, on the contrary, recommend among the numerous recent publications, the book by Roger Penrose, *The Emperor's New Mind* (Oxford, 1989), especially Chapter 7, "Cosmology and the Arrow of Time," where one can find an excellent resumé of the state of the question from the point of view of physical and cosmological theories.

The small book by P. C. W. Davies, *The Cosmic Blueprint* (London, 1987), may help to convince those who are in need of arguments that there is nothing in contemporary science allowing the continuation of the positivistic blackmail, which allows the (non-positivist) philosophers to continue in their dogmatic slumber.

The translations of the citations of Aristotle and Augustine are mine.

SMITH: *A Slip in Time Saves Nine*

1. The most accessible account of the ritual is in Livy, *History of Rome,* 7.3. Compare the interpretations of J. G. Frazer, *The Golden Bough,* 3rd ed. (New York, 1935), 9: 66–67 and n. 1; and G. Piccaluga, "La scrittura coercitive," *Cultura e Scuola,* 25 (1983): 117–24. I owe the latter reference to Mr. Rob Campany.

2. Note should be taken at the outset of the far more complex and supple typology of scholars such as G. Gurvitch, *The Spectrum of Social Time* (Dordrecht, 1964), pp. 13–14 et passim, who distinguishes "eight different kinds of social time as frames of reference for sociological analysis." I know of no comparable recognition of complexity within the field of religious studies.

3. In the last two paragraphs I have drawn on my essay "Mythos und Geschichte," in *Alcheringa oder die beginnende Zeit,* ed. H. P. Duerr (Frankfurt am Main, 1983), pp. 29–49, esp. pp. 35–41, which supplies the necessary documentation.

4. H. Gese, "The Idea of History in the Ancient Near East and in the Old Testament," *Journal for Theology and the Church,* 1 (1965): 45.

5. B. M. Metzger, "Methodology in the Study of the Mystery Religions and Early Christianity" (1955), reprinted in B. M. Metzger, *Historical and Literary Studies: Pagan, Christian, Jewish* (Grand Rapids, Mich., 1968), p. 23. Cf. his briefer version of the same statement in the *Twentieth Century Encyclopedia of Religious Knowledge* (1955), 2: 772.

6. C. Lévi-Strauss, *The Savage Mind* (Chicago, 1966), p. 95.

7. See E. Durkheim, *The Elementary Forms of the Religious Life* (reprint, New York, 1965), p. 391, who argues, in the case of ritual, that "the seasons have only furnished the outer framework (*le cadre extérieur*)" for what are, in his interpretation, actually social-communal rhythms. C. Lévi-Strauss, *The Origin of Table Manners* (New York, 1978), esp. pp. 89–131, offers a profound meditation on notions of periodicity and their relationships to seasonal models in myth.

8. For a review of the evidence and the basic bibliography, see J. Z. Smith, "Dying and Rising Gods," in *The Encyclopedia of Religion,* ed. M. Eliade (New York, 1987), 4: 521–27.

9. For a preliminary attempt to map out such a soteriology with respect to one of these deities, see S. Sfameni Gasparro, *Soteriology and Mystic Aspects of the Cult of Attis and Cybele* (Leiden, 1985). For a more general treatment, see J. Z. Smith, *Drudgery Divine: On the Comparison of Early Christianities and the Religions of Late Antiquity* (London and Chicago, 1990).

10. This argument is the special contribution of C. Colpe's important study, "Zur Mythologischen Struktur der Adonis-, Attis-, und Osiris-Überlieferungen," in *Festschrift W. von Soden* (Neukirchen, 1969), pp. 23–44.

11. See A. E. Jensen, *Myth and Cult Among Primitive Peoples* (Chicago, 1963), pp. 5–6, 66, 171–76, 194, and the comments in J. Z. Smith, *Imagining Religion: From Babylon to Jonestown* (Chicago, 1982), pp. 42–43.

12. Unless otherwise acknowledged, I have taken all the quotations in this section from M. Eliade, *The Sacred and the Profane: The Nature of Religion* (New York, 1959), pp. 20–24, 64, 68–113. For a full discussion, see J. Z. Smith, *Map Is Not Territory* (Leiden, 1978), pp. 91–96.

13. M. Eliade, *Patterns in Comparative Religion* (New York, 1958), p. 460.

14. M. Eliade, *Traité d'histoire des religions,* 2nd ed. (Paris, 1964), p. 312.

15. On this point, I turn to a personal anecdote. When Eliade was asked in a seminar by Paul Ricoeur what it was that he wanted one to do with myth, Eliade told a story (or was it a parable): "In Rumania, at the first sign of spring, the peasants tramp through the snow looking for the first flower that has broken through the crust. When they spy it, they point to it, with their mouths open in delight." This silent, inarticulate pointing in the face of an "irruption" seems precisely to encapsulate one aspect of Eliade's understanding of the appropriate human response to a hierophany.

16. Eliade, *Patterns,* pp. 29–30, 459.

17. In addition to the vocabulary of human creativity, Eliade has a quite different language of transcendence in which one "abolishes" one's humanity, or ascends to another sphere, or "breaks through the roof."

18. All quotations are from E. Durkheim, *Elementary Forms,* pp. 18, 57, 246–50. See further J. Z. Smith, *To Take Place: Toward Theory in Ritual* (Chicago, 1987), pp. 39–40.

19. M. Mauss and H. Beuchat, "Essai sur les variations saisonnières des sociétés eskimos," *Année Sociologique,* 9 (1906): 39–132. Cf. H. Hubert, "Etude sommaire de la représentation du temps dans la religion et la magie," *Annuaire de l'Ecole Pratique des Hautes Etudes,* Section des sciences religieuses (1909): 1–39.

20. See Smith, *To Take Place,* pp. 106–8.

21. Smith, *To Take Place,* pp. 109–10.

22. Ph. Melanchthon, *De Luthero et aetatibus ecclesiae* (1548), in *Corpus Reformatorum* (Halle, Brunswick, Berlin, 1834–), 11: 786.

GOODY: *The Time of Telling and the Telling of Time*

I am indebted to Bliss Carnochan for his remarks and to Cindy Ward for comments about dates in Robinson Crusoe; in addition, the participants at the conference pointed to areas where I needed to supplement the argument, which draws implicitly on my earlier writing as well as on current interests. Much necessarily remains allusive in a short paper on a general topic.

1. In the past my own writing about time, its concepts, its measurement, its organization, has been deliberately quotidian, prosaic. That is to say, I have tried to look critically at the great cultural generalizations—time in circles, time in straight lines, time autonomous, time embedded—or rather to look both critically and sideways at them from a more determinative point of view, asking what particular features of social organization, technology, or other aspects of cultural action would tend to influence concepts of time in certain directions. This I did because it seemed to me too easy, too facile, too literary perhaps, to try and characterize chronotypes in largely personal terms, throwing words into the air like confetti and expecting the colored fragments to fall into some enduring pattern. So my article in an encyclopedia on the social organization of time (*The International Encyclopedia of the Social Sciences,* 1968) looked at the relation of concepts of time to methods of measurement and assessment, in the broadest sense, as well as to modes of production, modes of communication, and to religious ideas about the world. I return to that theme, but what struck my attention in the present context was the insistent emphasis on time and narrative in written literature, for narrative has a very special relationship to time, being located at a specific conjunction in time and space, and unwinding in the framework of time.

2. I. P. Watt, *The Rise of the Novel* (London, 1957), pp. 15, 22, 21.

3. See J. Goody, *The Domestication of the Savage Mind* (Cambridge, Eng., 1977) and *The Logic of Writing and the Organisation of Society* (Cambridge, Eng., 1986).

4. Briefly, a narrative involves not simply sequence but a relation between the elements involving the same characters or possibly objects.

5. See C. M. Bowra, *Primitive Song* (London, 1962); Ruth Finnegan, *Oral Literature in Africa* (Oxford, 1970).

6. G. S. Kirk, *Myth: Its Meaning and Function in Ancient and Other Cultures* (Cambridge, Eng., 1970).

7. J. Goody, *The Interface Between the Written and the Oral* (Cambridge, Eng., 1987). See M. Parry, *The Making of Homeric Verse* (A. Parry, ed.; Oxford, 1971), A. B. Lord, *The Singer of Tales* (Cambridge, Mass., 1960), H. M. Chadwick, *The Heroic Age* (Cambridge, Eng., 1912), and H. M. and N. K. Chadwick, *The Growth of Literature* (3 vols.; Cambridge, Eng., 1932–40).

8. Finnegan, *Oral Literature*.

9. J. Goody, *The Myth of the Bagre* (Oxford, 1972), and J. Goody and S. W. D. K. Gandah, *Une Récitation du Bagre* (Paris, 1981).

10. Webster's dictionary defines plot as "the main story of a literary work . . . unfolding of a carefully connected sequence of motivated incidents."

11. S. H. Hooke, *Myth, Ritual and Kinship* (Oxford, 1960).

12. Contrary to F. A. Havelock's statement about the Bagre in *The Muse Learns to Write: Reflections on Orality and Literacy from Antiquity to the Present* (New Haven, Conn., 1986), pp. 28, 44, I see no evidence of Muslim influence, though I have long pondered the possibility.

13. See J. A. Braimah, *Gonja Drums* (Accra, Ghana, n.d.).

14. See I. Wilks and N. Levtzion, *Chronicles from Gonja* (Cambridge, Eng., 1987).

15. J. Goody, *The Ethnography of the Northern Territories of the Gold Coast, West of the White Volta* (Colonial Office, London, 1954).

16. M. Cole, J. Goody, and S. Scribner, "Writing and Formal Operations: A Case Study Among the Vai," reprinted in Goody, *Interface.*

17. On the Tiv, see L. Bohannan, "A Genealogical Charter," *Africa,* 22 (1952): 301–15.

18. J. Goody and I. P. Watt, "The Consequences of Literacy," *Comparative Studies in Society and History,* 5 (1963): 304–45; Goody, *Logic of Writing.*

19. See Peter Brooks, "The Storyteller," *The Yale Journal of Criticism,* 1 (1987): 21–38, for a perceptive account of what he calls "the oral in the written."

SPIVAK: *Time and Timing*

1. "Time is not an empirical concept that has been derived from any experience, for neither coexistence nor succession would ever come within our perception, if the representation of time were not the *a priori* grounding [*zum Grunde läge*]. . . . Time is a necessary representation which grounds [*zum Grunde liegt*] all intuition," Immanuel Kant, "Time," in *Critique of Pure Reason,* trans. Norman Kemp Smith (New York, 1965), p. 74. All translations, including this one, have been modified as I have considered necessary. I always use the shocking "he" when that is true to the spirit of the author. Kant's system cannot be made socio-sexually just by pronominal piety, without violating the argument. (See, for example, Genevieve Lloyd, *The Man of Reason: "Male" and "Female" in Western Philosophy* [Minneapolis, 1984], chap. 4.) This also reminds some of us, as we speculate about the ethics of sexual difference, that traditional European ethical philosophy simply disavows or benevolently naturalizes its sexual differentiation.

2. The best explanation of this argument is still Jacques Derrida, "Freud and the Scene of Writing," in his *Writing and Difference,* trans. Alan Bass (Chicago, 1978).

3. For the "good and bad writing" argument, see Jacques Derrida, *Of Grammatology,* trans. Gayatri Chakravorty Spivak (Baltimore, Md., 1976), pp. 15–18.

4. Gayatri Chakravorty Spivak, *Master Discourse, Native Informant: Deconstruction in the Service of Reading* (Cambridge, Mass., forthcoming).

5. Max Horkheimer said this powerfully over 50 years ago: "The attempt to afford justification to every idea and every historical person and to assign the heroes of past revolutions their place in the pantheon of history next to the victorious generals of the counterrevolution, this ostensibly free-floating objectivity conditioned by the bourgeoisie's stand on two fronts against absolutist restoration and against the proletariat, has acquired validity in the Hegelian system along with the idealistic pathos of absolute knowledge" ("On the Problem of Truth," in *The Essential Frankfurt School Reader,* ed. Andrew Arato and Eike Gebhardt [New York, 1978], p. 418).

6. G. W. F. Hegel, *The Philosophy of History,* trans. J. Sibree (New York, 1956); *The Philosophy of Right,* trans. T. M. Knox (Oxford, 1962); *Aesthetics:*

Lectures on Fine Arts, trans. T. M. Knox, 2 vols. (Oxford, 1975). All references to the last of the three are included in my text as *LA,* followed by page number.

7. G. W. F. Hegel, *The Philosophical Propaedeutic,* trans. A. V. Miller (Oxford, 1986), p. 126.

8. I have signaled a relationship between a political unconscious and the Hegelian epistemograph in my *In Other Worlds: Essays in Cultural Politics* (New York, 1987), pp. 258–59. Surprisingly, there is an acceptance of "normative deviations" in Fredric Jameson, *Marxism and Form: Twentieth-Century Dialectical Theories of Literature* (Princeton, N.J., 1971), pp. 329–30. It is, I think, this conviction or presupposition that surfaces in Jameson's by-now-notorious essay "Third World Literature in the Era of Multinational Capital," *Social Text,* 15 (Fall 1986): pp. 65–88, contested by Aijaz Ahmed, "Jameson's Rhetoric of Otherness and the 'National Allegory,'" *Social Text,* 17 (Fall 1987): pp. 3–27. The terms of the contestation are, in this reading, a questioning of the "scientific" claims of the Hegelian epistemograph, however disguised. For a brilliant "commonsense" interpretation that sees this characteristic yet misses its point, see G. A. Cohen's treatment of Hegel in *Karl Marx's Theory of History: A Defence* (Princeton, N.J., 1978).

9. What Hegel is producing and presupposing here is a semitized "orientalist," nearly monotheist, homogeneous religion called "Hinduism." See *Seminar* (Sept. 1985). A somewhat psychologistic account of the construction of this religion is also to be found in Ashis Nandy, *The Intimate Enemy: Loss and Recovery of Self Under Colonialism* (Delhi, 1983). "Taking brahmin documents as representative of all Indian society" is of course still common practice (Damodar Dharmanand Kosambi, *Myth and Reality: Studies in the Formation of Indian Culture* [Bombay, 1983], p. 38 n. 3). This approach is to be strictly distinguished from analyzing such documents as ingredients of a regulative psychobiography—as in my "Can the Subaltern Speak?" in *Marxism and the Interpretation of Culture,* ed. Larry Grossberg and Cary Nelson (Urbana, Ill., 1988), or in Gilles Deleuze and Felix Guattari, *Anti-Oedipus: Capitalism and Schizophrenia,* trans. Robert Hurley et al. (Minneapolis, Minn., 1986), p. 296. To confuse the two approaches is to be tendentious.

10. Hegel, *Philosophy of History,* p. 99. V. Y. Mudimbe, *The Invention of Africa: Gnosis, Philosophy, and the Order of Knowledge* (Bloomington, Ind., 1988), is, among many other things, a brilliant and judicious example of how to situate Hegel and other European "thinkers" of Africa. This book, supplemented by books such as Jack Forbes, *Black Africans and Native Americans: Color, Race and Caste in the Evolution of Red-Black People* (Oxford, 1988), and Roger D. Abrahams, John S. Szwed, Adrian Stackhouse, and Leslie Baker, eds., *After Africa: Extracts from British Travel Accounts and Journals of the Seventeenth, Eighteenth, and Nineteenth Centuries Concerning the Slaves, Their Manners, and Customs in the British West Indies* (New Haven, Conn., 1983), opens a beginning for the nonspecialist. For a general discussion of *Philosophy of History* on the matter of India, see Perry Anderson, *Lineages of the Absolutist State* (London, 1974), pp. 466–67.

11. As Knox notes, Zaehner has translated this last bit as follows: "By these three states of being inhering in the constituents the whole universe is led astray and does not understand that I am far beyond them and that I neither change nor pass away." This translation can also be questioned.

12. An argument concentrating on the logic of the metonym would point out that the figurative energy of the text pushes the "earlier" semiotic field of lineage into the "later" one of the nascent state (see Romila Thapar, *From Lineage to State: Social Formations in the Mid-First Millennium* B.C. *in the Ganga Valley* [Bombay, 1984]). In question here is the killing of blood kin, forbidden in the earlier formation. Krishna himself might be a mark of "the transition from mother-right to patriarchal life, [which] allowed the original cults to be practised on a subordinate level" (Kosambi, *Myth,* p. 28). Part of the figurative logic might be based on the possible regulative norm of sanctioned suicide (it is allowed to kill and be killed when you know that the soul is immortal), of which I have written elsewhere (Spivak, "Can the Subaltern Speak?"). Whatever else it might be, it is not a *monotonous* argument.

13. Some of B. K. Matilal's titles are "Images of India: Problems and Perceptions," "In Defence of Devious Krsna," "Epic and History: Tradition, Dissent and Politics in India," "Caste, Karma and the Gita," "Moral Dilemmas: Insights from Indian Epics." There are plans to incorporate these in a volume on ethics and Indian cultural studies.

14. Kosambi, *Myth,* p. 15.

15. Ibid., p. 17.

16. *The Bhagavadgita in the Mahabharata,* trans. J. A. B. van Buitenen (Chicago, 1981), is the definitive bilingual edition accessible to the nonspecialist. I have offered my own translations of the Sanskrit because often this is the only way to cut through the solemnity that informs the best translations of the great texts of classical antiquity. I have tried to follow the contemporary phonetic transcriptions of Sanskrit words, except with such words as "Sanskrit" or "Krishna," which are well known to the nonspecialist reader in these nonspecialized spellings. I apologize for the inconsistency but, since this is recognizably not an expert Indianist text, I felt that consistency would have been an affectation.

17. See, for example, Sarvepalli Radhakrishnan, *The Hindu View of Life* (London, 1961), p. 18, from which I will quote at greater length later.

18. Hegel's traffic with India is ably criticized by Michel Hulin, *Hegel et l'orient: Suivi de la traduction annotée d'un essai de Hegel sur le Bhagavad-Gita* (Paris, 1979). Hulin includes Hegel's two reviews on the subject of the *Gita* and on its relationship to the philosophy of India. Any serious consideration of the specific topic of Hegel's orientalism would have to examine these essays in detail. My interest is in noticing how the well-known texts are woven with the axiomatics of imperialism, and therefore I keep to the *LA*. It is clear from Hegel's letters that he was well acquainted with contemporary German scholars of Sanskrit. I have not yet read Du-Yol Song, *Die Bedeutung der asiatischen Welt bei Hegel, Marx und Max Weber,* a dissertation written under Jürgen Habermas in 1972. I use "preconscious" to distinguish Indian from Persian art in the Hegelian morphology. The "luminous essence" that provides the pre-originary space for the Persian scene of fire has been delicately discussed by Werner Hamacher (unpublished lecture, Stanford University, May 10, 1988). I might mention in passing that, although I am deeply interested in the usual deconstructive focus (not always shared by Derrida) on the "moments" (I use this word where no word will suffice) of "stalling" (Hamacher's word) at beginning and end ("differance" and "aporia" are

only two names for these moments), I am more interested in the generating of a shaky middle by way of an irreducible "mistake" (not to be derived from some prior "correct" step). I have touched on this issue in "Feminism and Deconstruction, Again," in *Between Feminism and Psychoanalysis,* ed. Teresa Brennan (London, 1989). The reading of *Antigone* is in *The Phenomenology of Mind,* trans. A. V. Miller (Oxford, 1977), pp. 261–62, 284–89. I will take the liberty of a digression here. Just as Hegel telescopes 2,500 years to prove that Indians cannot move history, so also does he base his evidence for the Indian "recurring description of natural *generation* instead of the idea of a spiritual *creation*" on absent passages that he could not have read: "(This passage the English translator had no mind to translate word for word because it is all too wanting in decency and shame.) . . . Schlegel has not translated this part of the episode" (*LA* 1: 344).

19. Most strikingly in Jacques Derrida, *Glas,* trans. John P. Leavey, Jr., and Richard Rand (Lincoln, Nebr., 1986).

20. I have discussed this perspective briefly in "Imperialism and Sexual Difference," in *The Current in Criticism: Essays on the Present and Future of Literary Theory,* ed. Clayton Koelb and Virgil Lokke (West Lafayette, Ind., 1988), and more elaborately in my forthcoming book, *Master Discourse, Native Informant*. For the value of the "(im)," see Spivak, *In Other Worlds,* p. 263. Such positions breed a new politics of representing and reading the past. In one way or another, many resistant postcolonial readers are coming to imagine such positions. I read with a degree of solidarity "the anthropoetical effort" of Jose Piedra: "There must be anthropoetical traces left in the colonial text to show for the underlying inequalities and incompetences of the game of arrival. These traces . . . belong to the native and represent the creole learnings of critics" ("The Game of Critical Arrival," *Diacritics,* Spring 1989, p. 191).

21. Kosambi, *Myth,* p. 18.

22. "Unchanging way of knowing" is *avyaya yoga* in the original, meaning something more like "undiminishing technique." The problem of translating *yoga* is well known. It might interest the reader to know that the qualifier for the noun is the participial nominative for the indeclinable particle.

23. My "native informant" should be able to think Derrida if Derrida's notion of good divine writing as set up against bad human writing, of which male auto-affection is a case, has any logical plausibility. I hope the reader is able to distinguish this suggestion from the appropriation of "deconstruction" to legitimize exotic texts. As far as I can tell, there is no recriminatory pathos against masturbation in "Hindu" regulative psychobiography. For the authoritative injunction against masturbation, see *The Laws of Manu,* trans. G. Buhler (Oxford, 1886), canto 2, lines 180–81, p. 63. I am grateful to Bimal Krishna Matilal for this reference.

24. My own bilingual copy of the *Gita,* purchased in my teens when I was profoundly taken by nineteenth-century semitized Hinduism, offers the standard conceptual-anagogical reading within that framework: "Though I am unborn, of changeless nature and Lord of beings, yet subjugating My Prakriti, I come into being by My own Maya" (*Shrimad-Bhagavad-Gita,* trans. Swami Swarupananda [Calcutta, 1956], p. 99). This translation is perhaps more marked by the West than a reading using Greek words carrying *for the moment* a "Derridean" flavor. Van Buitenen's authoritative translation is: "Although indeed I am unborn and

imperishable, although I am the lord of the creatures, I do resort to nature, which is mine, and take on birth by my own wizardry" (p. 87).

25. Much can be made of the fact that *darṣana*—vision—is usually translated as "philosophy," although that usage would be clearly inappropriate here. The alternative usage, the felicity of constituting the transcendental object as object of the gaze, at once points at the difficulty of violating a cultural text by translation (*darṣana* = philosophy = idolatrous reverence; ergo India [= Sanskritic Hinduism] has no philosophy but only religion/superstition), and offers the possibility of a deconstructive lever on the model of *pharmakon* (drug/poison), supplement (addition/hole-filler), *differance* (effacement/disclosure), and the like. (Can the "same thing" be done with *theorein*—to see? I do not know.) Since I am not offering a deconstructive reading of the *Gita* here, but rather using deconstruction as an excuse for the figuration of my perspective, I have no interest in pursuing this line of speculation. I should also mention that the reward-for-human-frailty *topos* is used abundantly in the *Gita* and elsewhere as the legitimization of *bhaktiyoga*, commonly translated as "the path of devotion." Kosambi points out that "to hold feudal [this term is now contested by Indian historians] society and its state together, the best religion is one which emphasizes the role of *bhakti*, personal faith, even though the object of devotion may have clearly visible flaws" (Kosambi, *Myth*, p. 32). As the following pages will, I hope, make clear, this is different from my analysis of the foregrounding of human error as the rhetorical motor of a shift from the transcendental to the social.

26. Arjuna is speaking here of the difference between human and divine. With all due respect, it seems to me to be merely pedantic to explain this by reference to mere "matters of [social] precedence," as does van Buitenen (p. 167 n. 9). Kosambi's down-to-earth approach necessarily flattens the text, for it does not read closely: "the moral is pointed by the demoniac God himself: that all the warriors on the field had really been destroyed by him; Arjuna's killing them would be a purely formal affair whereby he could win the opulent kingdom" (*Myth*, p. 17).

27. Thomas A. Sebeok, *Style in Language* (Cambridge, Mass., 1960), p. 358.

28. In the extended study of which this essay is a part, I draw a parallel here to the comparative *general* neglect of Kant's "Critique of *Teleological* Reason," in his *Critique of Judgment*.

29. Hitherto only the proper name *ksatriya* (warrior) is used a number of times, not as one caste marker among four, but as a general interpellation in ideology for Arjuna. The enunciative strategy of the verse (4.13) that is invariably chosen, by Radhakrishnan and many others, as proof of the liberating and flexible vocational definition of caste offered by Krishna in the *Gita* (Radhakrishnan, *Hindu View*, p. 79), should be carefully analyzed before any claim is made. Matilal locates a critical tradition within Brahmanical orthodoxy itself, however defined. His treatment of the field of battle as the field of *dharma* (*dharmakshetra*) and this, in its turn, as a field of rule-following; his critique of Max Weber by way of a commentary on the relationship between caste and Karma; and his astute tracking of the slippage between *svadharma* and *svabhāva* all relate his study, as I have already suggested, to the study of the formation of culture on the Indian subcontinent and its diasporic and global variants today. Hegel's general position on caste is to be found in *The Philosophy of History*, p. 168.

30. It would be interesting to work this into the alliance-affinal dialectic of territorialization and coding in Deleuze/Guattari, *Anti-Oedipus: Capitalism and Schizophrenia,* trans. Robert Hurley et al. (Minneapolis, Minn., 1983), pp. 145–46. Let us however remember that, although they are critical of the connections between ethnography and psychoanalysis, they themselves share some of the historical prejudices (such as a faith in "oriental despotism") sustained by the culture of imperialism. See, for example, Deleuze/Guattari, *A Thousand Plateaus: Capitalism and Schizophrenia,* trans. Brian Massumi (Minneapolis, Minn., 1987), pp. 351–52.

31. Spivak, "Can the Subaltern Speak?" The contrast between Matilal's notion of Krishna as emulable moral agent and this idea of Krishna as exceptionalist regulator nicely points up the difference between analytic and deconstructive studies of culture.

32. Karl Marx, *Early Writings,* trans. Rodney Livingstone and Gregor Benton (Harmondsworth, Eng., 1975), p. 393.

33. Sigmund Freud, *The Standard Edition of the Complete Psychological Works,* trans. James Strachey et al. (London, 1961), 19: 236.

34. Marx, *Early Writings,* p. 324; emphasis Marx's. This is not the place to pursue my conviction that *Entfremdung* (estrangement) and *Entäusserung* (alienation) generally carry separate charges in Marx—the first is an ontological error perpetrated by philosophy in collaboration with political economy, the second an ontological necessity for the very predication of (human) being and doing.

35. Partha Chatterjee, *The Nationalist Resolution of the Women's Question,* Centre for Studies in Social Sciences, Occasional Paper no. 94 (Calcutta, 1987), p. 6. See also Kumari Jayawardena, *Feminism and Nationalism in the Third World* (London, 1986), pp. 254–61 and passim.

36. Nandy, *Intimate Enemy,* p. 47.

37. Aurobindo Ghose, *Essays on the Gita* (Pondicherry, 1974); Radhakrishnan, *The Hindu View* (cited in n. 17); S. G. Sardesai and Dilip Bose, *Marxism and the Bhagvat Geeta* (New Delhi, 1982).

38. Ghose, *Essays,* p. 3.

39. Radhakrishnan, *Hindu View,* pp. 32–33.

40. Karl Marx, *Surveys from Exile,* trans. David Fernbach (New York, 1974), p. 146.

41. Sardesai and Bose, *Marxism,* p. 24.

42. "Internationalism of Oppressors," *Economic and Political Weekly,* Jan. 23, 1988, p. 108.

43. Julia Kristeva, "My Memory's Hyperbole," in *The Female Autograph: Theory and Practice of Autobiography from the Tenth to the Twentieth Century,* ed. Domna C. Stanton (Chicago, 1987), p. 235.

44. This type of assertion provokes resentment in metropolitan anticolonialists as well as a growing body of indigenous urban intellectuals who are themselves critical of hegemonic nationalism in India and yet are paradoxically susceptible to identifying "India" with the view from the urban centers where they live and teach. This can, on occasion, turn into a rather insidious brand of nationalism disqualifying all diasporic analysis. Without prejudice to the further development of an argument analyzing this position, I should like to offer as appeasement the assurance that my assertion is also about the nature of "real" origins in general, and

works against the authority of abundantly established dominant origins such as "Hegel." Our vigilance relates to the counterclaims to alternative origins, expressed by the indigenous dominant as self-chosen representative of the subordinate, legitimizing the vanguardism of established "origins" by a mere reversal.

LA CAPRA: *The Temporality of Rhetoric*

A slightly different version of this paper appears as Chapter 4 of my *Soundings in Critical Theory* (Ithaca, N.Y., 1989). I am grateful to the publisher, Cornell University Press, for permission to use the paper here.

1. M. H. Abrams, *Natural Supernaturalism* (New York, 1973). All references will be to this edition, and page numbers will be included in parentheses in the text. The work was first published in 1971.

2. Paul de Man, "The Rhetoric of Temporality," in *Interpretation: Theory and Practice,* ed. Charles Singleton (Baltimore, Md., 1969). All references will be to this edition, and page numbers will be included in parentheses in the text. (The essay also appears in Paul de Man, *Blindness and Insight,* 2nd rev. ed. [Minneapolis, Minn., 1983].)

3. De Man does refer a number of times to Abrams's "Structure and Style in the Greater Romantic Lyric," in *From Sensibility to Romanticism: Essays Presented to F. A. Pottle,* ed. F. W. Hillis and H. Bloom (New York, 1965).

4. De Man's own self-commentary in the foreword to his revised, 1983 edition of *Blindness and Insight* (p. xii) is interesting in this regard: "With deliberate emphasis on rhetorical terminology, ["The Rhetoric of Temporality"] augurs what seemed to me to be a change, not only in terminology and in tone but in substance. This terminology is still uncomfortably intertwined with the thematic vocabulary of consciousness and of temporality that was current at the time, but it signals a turn that, at least for me, has proven productive. . . . When one imagines to have felt the exhilaration of renewal, one is certainly the last to know whether such a change actually took place or whether one is just restating, in a slightly different mode, earlier and unresolved obsessions." One may of course also raise the question of the extent to which de Man, in elaborating a theory of Romanticism in "The Rhetoric of Temporality," rewrites and in certain ways transforms Walter Benjamin's *Ursprung des deutschen Trauerspiels* (Berlin, 1928; rev. ed., Frankfurt, 1963), published in English as *The Origin of German Tragic Drama,* trans. John Osborne (London, 1977). The first footnote in "The Rhetoric of Temporality" would also indicate a role for Michel Foucault's conception of "epistemological breaks" as elaborated in *Les Mots et les choses* (Paris, 1966), published in English as *The Order of Things* (New York, 1970). De Man is, however, less concerned with breaks between periods than with breaks or radical disjunctions within periods that problematize the very conception of periodization.

5. Baudelaire, *De l'essence du rire,* in his *Oeuvres complètes* (Paris, 1968), p. 378. All references are to this edition, and page numbers will be included in parentheses in the text. Translations are my own.

6. It is strange that de Man, in the original edition of his essay, refers repeatedly to E. T. H. Hoffmann. This mistake is particularly bizarre in that someone of de Man's erudition must have known that Hoffmann chose the name

Amadeus to mark his respect and love for Mozart. (Hoffmann's given name was Ernst Theodor Wilhelm.)

7. Actually these caricatures are Hoffmann's revisions of Callot's drawings. The latter do have realistic street scenes as settings. In Hoffmann, the stock figures are set on an abstracted piece of cloudlike turf.

8. A problematic distinction may be drawn between temporality and historicity, but not a clear-cut opposition or dichotomy. Temporality refers to the movement of time in general. Historicity involves the variable insertions of subjects in the temporal process as historical agents and historiographers. Historicity is itself made possible in and through signifying practices, and it is on the basis of such practices that one may ascribe temporality to other processes. Historicity also raises the issue of how to "represent" temporal processes that bear on subjects without being restricted to (or even centered on) them—processes related to the articulation and disarticulation of subject positions.

9. See, for example, Peter Brooks, Shoshana Felman, and J. Hillis Miller, eds., *The Lesson of Paul de Man, Yale French Studies,* 69 (1985); Stanley Corngold, "Error in Paul de Man," in *The Yale Critics: Deconstruction in America,* ed. Jonathan Arac, Wlad Godzich, and Wallace Martin (Minneapolis, Minn., 1983); Jacques Derrida, *Memoires for Paul de Man* (New York, 1986); Rodolphe Gasché, "*Setzung* und *Übersetzung:* Notes on Paul de Man," *Diacritics,* 11 (Winter 1981): 36–57; Suzanne Gearhart, "Philosophy Before Literature: Deconstruction, Historicity, and the Work of Paul de Man," *Diacritics,* 13 (Winter 1983): 63–81; Lindsay Waters and Wlad Godzich, eds., *Reading de Man Reading* (Minneapolis, Minn., 1988). See also Werner Hamacher, Neil Hertz, and Thomas Keenan, eds., *Responses: On Paul de Man's Wartime Journalism* (Lincoln, Nebr., 1989). Except for a few revisions, the present essay was written before the discovery of de Man's World War II journalism, which poses problems in interpretation that I do not address.

LUCKMANN: *The Constitution of Human Life in Time*

1. This is a point that I have argued more extensively elsewhere. Cf. Thomas Luckmann, "Personal Identity as an Evolutionary and Historical Problem," in *Human Ethology: Claims and Limits of a New Discipline,* ed. Mario von Cranach, Klaus Foppa, Wolf Lepenies, and Dieter Ploog (Cambridge, Eng., 1979), pp. 56–74; and "Remarks on Personal Identity: Inner, Social and Historical Time," in *Identity: Personal and Socio-Cultural,* ed. Anita Jacobson-Widding, Acta Universitatis Upsaliensis, Uppsala Studies in Cultural Anthropology, no. 5 (Uppsala, 1984), pp. 67–91. The remarks and observations that follow are based to a considerable extent on analyses first presented in the two contributions just mentioned, as well as upon two partly overlapping essays that deal directly with the problem of time: Thomas Luckmann, "Lebensweltliche Zeitkategorien, Zeitstrukturen des Alltags und der Ort des 'historischen Bewusstseins,'" in *Grundriss der romanischen Literaturen des Mittelalters,* vol. 11, no. 1, ed. Hans Robert Jauss et al. (Heidelberg, 1986), pp. 117–26, and "Gelebte Zeiten—und deren Ueberschneidungen im Tages- und Lebenslauf," in *Epochenschwelle und Epochenbewusst.*

Poetik und Hermeneutik XII, ed. Reinhart Herzog and Reinhart Koselleck (Munich, 1987), pp. 283–304.

2. Kurt Lüscher, "Time, a Much Neglected Dimension in Social Theory and Research," *Sociological Analysis and Theory,* 4 (1974): 101–17.

3. Maurice Halbwachs, *Les Cadres sociaux de la mémoire* (Paris, 1925), and *La Mémoire collective* (Paris, 1950), published in English as *The Collective Memory* (New York, 1980).

4. Cf. especially Georg Simmel, "Das Problem der historischen Zeit," in *Brücke und Tür* (Stuttgart, 1957), pp. 43–58.

5. George Herbert Mead, *The Philosophy of the Act* (Chicago, 1938), and *Mind, Self, and Society* (Chicago, 1934).

6. Alfred Schutz, *Der sinnhafte Aufbau der sozialen Welt* (Vienna, 1932), published in English as *The Phenomenology of the Social World,* trans. G. Walsh and F. Lehnert (Evanston, Ill., 1970); idem, "Common Sense and the Scientific Interpretation of Human Action," in *Collected Papers I: The Problem of Social Reality* (The Hague, 1962), pp. 3–47; idem, "Tiresias, or Our Knowledge of Future Events," in *Collected Papers II: Studies in Social Theory* (The Hague, 1964), pp. 277–93; Alfred Schutz and Thomas Luckmann, *The Structures of the Life-World I* (Evanston, Ill., 1973); idem, *Strukturen der Lebenswelt II* (Frankfurt, 1984).

7. Norbert Elias, *Über den Prozess der Zivilisation* (Basel, 1939), published in English as *The Civilizing Process* (New York, 1978).

8. Only a few instances of significant work in this area are Barney G. Glaser and Anselm L. Strauss, *Status Passage* (London, 1971); Wilbert E. Moore, *Man, Time and Society* (New York, 1963); Julius Roth, *Timetables* (New York, 1963); Eviatar Zerubavel, *Hidden Rhythms: Schedules and Calendars in Social Life* (Chicago, 1981).

9. For a review of the state of theory and research on time in sociology, with a bibliography and additional references to significant theoretical contributions by Pitirim Sorokin, Robert Merton, Georges Gurvitch, Niklas Luhmann, and others see Werner Bergmann, "Das Problem der Zeit in der Soziologie," *Kölner Zeitschrift für Soziologie und Sozialpsychologie,* 35, no. 3 (1983): 462–504.

10. Schutz, "Common Sense," pp. 3–47, esp. pp. 19ff. Alfred Schutz, "Choosing Among Projects of Action," in *Collected Papers I,* pp. 67–96, esp. p. 67. Schutz and Luckmann, *Structures.*

11. Edmund Husserl, *Zur Phänomenologie des inneren Zeitbewusstseins (1893–1917),* vol. 10 of *Husserliana* (The Hague, 1966), esp. pp. 76 ff. See also Ludwig Landgrebe, "Phänomenologische Analyse und Dialektik," *Phänomenologische Forschungen,* 10 (1980): 21–88, esp. 66 ff.

12. Edmund Husserl, *Erfahrung und Urteil* (Hamburg, 1954), esp. secs. 17 ff.; idem, *Ideen zu einer reinen Phänomenologie und phänomenologischen Philosophie,* vol. 3, no. 1 of *Husserliana* (The Hague, 1950), esp. secs. 27 ff.; idem, *Die Krisis der europäischen Wissenschaften und die transzendentale Phänomenologie—Eine Einleitung in die phänomenologische Philosophie,* vol. 6 of *Husserliana* (The Hague, 1962), esp. sec. 37.

13. William James, *The Principles of Psychology,* vol. 1 (New York, 1890), esp. chap. 1.

14. For a detailed phenomenological investigation of synchronization see Schutz, *Der sinnhafte Aufbau,* esp. chap. 3.

15. It seems to be established that synchronization occurs very early in the life of infants. See C. Trevarthen, "Instincts for Human Understanding and for Cultural Cooperation: Their Development in Infancy," in *Human Ethology,* ed. Cranach, Foppa, Lepenies and Ploog, pp. 530–71.

16. See the unpublished dissertation of Devorah Kalekin-Fishman, "Time, Sound and Control: Aspects of Socialization in the Kindergarten" (University of Konstanz, 1980), for an interpretation of "musical instruction" in the broadest sense in teaching time-coordination and management in the kindergarten.

17. For a summary of the history of time-notation see E. R. Leach, "Primitive Time-Reckoning," in *A History of Technology,* vol. 1, ed. Ch. Singer, E. J. Holmyard, and A. R. Hall (Oxford, 1954), pp. 110–27, and for a systematic treatment P. Janich, *Protophysik der Zeit* (Frankfurt am Main, 1980).

18. See, for example, R. A. Fothergill, *Private Chronicles: A Study of English Diaries* (London, 1974).

19. See Thomas Luckmann, "Grundformen der gesellschaftlichen Vermittlung des Wissens: Kommunikative Gattungen," in *Kultur und Gesellschaft,* ed. F. Neidhardt, M. Lepsius, and J. Weiss, special issue of *Kölner Zeitschrift für Soziologie und Sozialpsychologie,* 27 (1986): 141–211.

20. See Hans Blumenberg, *Lebenszeit und Weltzeit* (Frankfurt am Main, 1986).

21. For a usefully balanced early contribution on historical consciousness in preliterate societies see Rüdiger Schott, "Das Geschichtsbewusstsein schriftloser Völker," *Archiv für Begriffsgeschichte,* 12 (1968): 166–205. A particularly illuminating study of the conceptions of time involved in elementary historical consciousness in a mixed oral/literate culture can be found in J. Duggan's "The Experience of Time as a Fundamental Element of the Stock of Knowledge in Medieval Society," in *Grundriss der romanischen Literaturen des Mittelalters,* vol. 11, no. 1, ed. H. R. Jauss et al. (Heidelberg, 1987), pp. 127–34.

22. See Henri Frankfort, *Kingship and the Gods* (Chicago, 1948); James Breasted, *The Dawn of Conscience* (New York, 1933); Jan Assmann, *Aegypten: Theologie und Frömmigkeit einer frühen Hochkultur* (Stuttgart, 1984).

23. See Karl Löwith, *Von Hegel bis Nietzsche* (Zürich, 1941); Hans Barth, *Wahrheit und Ideologie* (Zürich, 1945).

24. Michel de Montaigne, *Essais III 10* (Paris, 1906–31).

25. Blumenberg, *Lebenszeit.*

HAREVEN: *Synchronizing Individual, Family, and Historical Time*

The research for this paper was supported by grants number 5R01AG02468 and 5R01AG06441 from the National Institute on Aging. The Japanese-U.S. comparisons were supported by the Social Science Research Council, the Japan Society for the Promotion of Science, and the U.S.-Japan Friendship Commission. I am grateful to Professor Kanji Masaoka for his collaboration on the U.S.-Japanese comparisons and to Professor Kiyomi Morioka for his leadership in this comparative project.

1. E. P. Thompson, *The Making of the English Working Class* (N.Y., 1963).

2. Tamara K. Hareven, *Family Time and Industrial Time* (Cambridge, Eng., 1982).

3. Tamara K. Hareven, ed., *Transitions: The Family and the Life Course in Historical Perspective* (New York, 1978); Glen Elder, "Family History and the Life Course," *Journal of Family History,* 2 (1979): 279–304.

4. Bernice L. Neugarten and Gunhild O. Hagestadt, "Age and the Life Course," in *Handbook of Aging and the Social Sciences,* ed. R. H. Binstock and E. Shanas (New York, 1976), pp. 35–56.

5. Tamara K. Hareven, "Family Time and Historical Time," *Daedalus,* 106 (1977): 57–70.

6. Dennis P. Hogan and Takashi Mochizuki, "Demographic Transitions and the Life Course: Lessons from Japanese and American Comparisons," *Journal of Family History,* 13 (1988): 291–305.

7. John Modell, Frank Furstenberg, and T. Hershberg, "Social Change and Transitions to Adulthood in Historical Perspective," *Journal of Family History,* 1 (1976): 7–32.

8. Glen Elder, "Historical Changes in Life Patterns and Personality," in *Life Span Development and Behavior,* vol. 2, ed. P. B. Baltes and O. G. Brim, Jr. (New York, 1979) pp. 117–59.

9. Elder, "Historical Changes."

10. Hareven, ed., *Transitions;* Hareven, *Family Time and Industrial Time.*

11. Glen Elder, *Children of the Great Depression: Social Change in Life Experience* (Chicago, 1974).

12. Hareven, *Family Time and Industrial Time.*

13. Peter Uhlenberg, "Cohort Variation in Family Life Cycle Experience of U.S. Females," *Journal of Marriage and the Family,* 35 (1974): 289–92.

14. Joseph Kett, *Rites of Passage: Adolescence in America, 1790 to the Present* (New York, 1977).

15. Bertram J. Cohler, "Personal Narrative and the Life Course," in *Life Span Development and Behavior,* vol. 4, ed. P. B. Baltes and O. G. Brim, Jr. (New York, 1982), pp. 205–41; Kanji Masaoka et al., "Turning Points: A Study in Qualitative Change," in *Family and Life Course of Middle-Aged Men,* ed. K. Morioka (Tokyo, 1985), pp. 58–107.

16. Elder, "Historical Changes."

17. Hareven, *Family Time and Industrial Time.*

18. We gathered the data on the Manchester children's cohorts by conducting extensive life history interviews of all the children of the historical cohort whom we were able to trace in Manchester and in other parts of the United States, and who agreed to be interviewed. We also interviewed the children's spouses and the siblings of the spouses. In addition to the qualitative textual analysis of the interviews, we also reconstructed individual migration, educational, and work histories for each individual in the two cohorts from the interview data, as well as from city directories, employee files, and vital records. In Shizuoka, the Japanese researchers carried out a series of interviews between 1981 and 1983, using survey questionnaires. The last interview stage in Shizuoka in 1984 was an intensive, open-ended one, in which the men were encouraged to describe their turning points in detail.

19. Tamara K. Hareven and Kanji Masaoka, "Turning Points and Translations: A Comparison of the Life Course of Several Cohorts in Japan and the United States," *Journal of Family History,* 13, no. 3 (1988): 271–89.

20. Hareven, *Family Time and Industrial Time.*

21. Glen Elder and Tamara K. Hareven, "Rising Above Life's Disadvantage: From Great Depression to World War," in *Children of Their Times: Developmental and Historical Insight,* ed. Glen H. Elder, Jr., John Modell, and Ross Parke (forthcoming).

22. Hareven and Masaoka, "Turning Points and Transitions." I am indebted to Kanji Masaoka for his collaboration on the comparative study of the life course of American and Japanese men.

23. Peter Uhlenberg, "Changing Configurations of the Life Course," in *Transitions,* ed. Hareven, pp. 66–97.

24. Hareven, *Family Time and Industrial Time;* Howard Chudacoff and Tamara K. Hareven, "From the Empty Nest to Family Dissolution: Life Course Transitions into Old Age," *Journal of Family History,* 4 (1979): 69–83.

25. Modell, Furstenberg, and Hershberg, "Social Change."

26. Chudacoff and Hareven, "Empty Nest."

27. Kiyomi Morioka, *Family and Life Course of Middle-Aged Men* (Tokyo, 1985).

FABIAN: *Of Dogs Alive, Birds Dead, and Time to Tell a Story*

1. Some years ago my thoughts and findings on that subject were published in Johannes Fabian, *Time and the Other: How Anthropology Makes Its Object* (New York, 1983). See also Johannes Fabian, "Rules and Process: Thoughts on Ethnography as Communication," *Philosophy of the Social Sciences,* 9 (1979): 1–26; and Johannes Fabian, "Culture, Time, and the Other," *Berkshire Review,* 20 (1985): 7–23.

2. I am speaking of cultural anthropology or social anthropology. Prehistory and archeology, which in the United States are closely linked with anthropology, have different problems with time. In their research they do not interact with living peoples. But living peoples are often involved in methodological comparison and always as the addressees of prehistorical discourse. We shall return to this later on.

3. Two authors whose work I encountered more recently are Reinhart Koselleck, *Vergangene Zukunft. Zur Semantik geschichtlicher Zeiten* (Frankfurt, 1984), translated by Keith Tribe as *Futures Past: On the Semantics of Historical Time* (Cambridge, Mass., 1985); and Krzysztof Pomian, *L'Ordre du temps* (Paris, 1984). I should perhaps be embarrassed by my failure, in *Time and the Other,* to notice Emmanuel Levinas's *Le Temps et l'autre* (Paris, 1979), a re-edition of lectures that originally appeared in 1945–46, now also translated by Richard A. Cohen as *Time and the Other* (Pittsburgh, Penn., 1987). Although I might have convinced the publisher to stick to the title I originally proposed for my book (*Anthropology and the Politics of Time*), I don't worry much about priority. As far as I can follow Levinas, I realize that we share some concerns. But we deal with different things under the same title. I am searching for anthropology's Other; Levinas raises questions that lead him to the ultimate Other. I try to understand

the politics of time; Levinas explores transcendence in the phenomenology of time experience. There is at least one eminent anthropologist who feels that in their concern for an Other anthropology and theology converge; see Mary Douglas, "The Hotel Kwilu—A Model of Models," *American Anthropologist,* 91 (1989): 855–65. I agree, although my reasons are not hers. To me, some common ground is required only inasmuch as anthropology is emancipation from theology.

4. Such an illuminating episode was the core of a paper in which I worked out the idea of shared time in ethnographic research; see Fabian, "Rules and Process."

5. See Johannes Fabian, "The Missing Text," in idem, *Power and Performance: Ethnographic Explorations Through Proverbial Wisdom and Theater in Shaba (Zaire)* (Madison, Wis., 1990), chap. 5.

6. See Dennis Tedlock, *The Spoken Word and the Work of Interpretation* (Philadelphia, 1983).

7. Examples abound from cultures all over the world of conceptualizing the other as beast or God. This is just to remind ourselves that we are here examining but one form of "othering."

8. Stephen Jay Gould, *Time's Arrow, Time's Cycle* (Cambridge, Mass., 1987); see also Fabian, *Time and the Other,* chap. 1.

9. See J. W. Burrow, *Evolution and Society: A Study in Victorian Social Theory* (Cambridge, Eng., 1966); George W. Stocking, *Race, Culture, and Evolution: Essays in the History of Anthropology* (New York, 1968).

10. Once again, we limit our argument to living "others"; that our relations to others' societies of the past may have to be rethought in the light of ideas discussed here was pointed out above.

11. R. R. Marett, *Anthropology* (London, 1925), pp. 185–87; this work was originally published in 1912.

12. On Radcliffe-Brown's solution for the problem of survivals see George W. Stocking, "Radcliffe-Brown and British Social Anthropology," in *Functionalism Historicized: Essays on British Social Anthropology,* ed. G. W. Stocking, vol. 2 of *History of Anthropology* (Madison, Wis., 1984), pp. 131–91.

13. Marvin Harris, *Culture, People, Nature: An Introduction to General Anthropology,* 2nd ed. (New York, 1975), p. 153.

14. The expression is taken here from the title of Koselleck's collection of essays. He provides references to other authors who have used the phrase; see Koselleck, *Vergangene Zukunft,* p. 17.

15. Nor is this the place properly to express my respect for attempts to rethink evolution with the help of much more sophisticated conceptions of time than were available when cultural evolutionism was first formulated. See Tim Ingold, *Evolution and Social Life* (Cambridge, Eng., 1986). Ingold, for instance, rejects the "time-machine" aspect of evolutionism and states: "*We* cannot move back in real time. . . . Yet if the Other is automatically defined as a predecessor, we would have to be able to do just that in order to participate with him. The fact that we can, nevertheless, participate with people wherever they may live, serves once again to demonstrate that mankind is advancing on a continuous front, rather like the crest of a spherical wave, and is not divided into ever so many discrete societies each following behind the other in lockstep" (p. 150).

16. See Bennetta Jules-Rosette, "The Veil of Objectivity: Prophecy, Divination and Social Inquiry," *American Anthropologist,* 80 (1978): 549–70; Carlo Ginzburg, "Morelli, Freud and Sherlock Holmes: Clues and Scientific Method," *History Workshop,* 9 (1980): 5–36; J. P. Vernant, L. Vandermeersch, J. Gernet, L. Bottéro, R. Crahay, L. Brisson, J. Carlier, D. Grodzynski and A. Retel-Laurentin, *Divination et rationalité* (Paris, 1974).

17. James Clifford and George E. Marcus, eds., *Writing Culture: The Poetics and Politics of Ethnography* (Berkeley, Calif., 1986); see also Clifford Geertz, *Works and Lives: The Anthropologist as Author* (Stanford, Calif., 1988).

18. I have found much food for thought about narrativity in a special issue of *Critical Inquiry* on the topic, with contributions by, among others, P. Ricoeur and V. Turner, and especially H. White, who gives a concise and cogent analysis of the ontological and political foundations of narrative in our Western tradition, of which anthropology is, of course, a part; see Hayden White, "The Value of Narrativity in the Representation of Reality," *Critical Inquiry,* 7 (1980): 5–27. For a thoughtful treatment of the problem of narration and knowledge (in a philosophical perspective I do not fully share) see Arthur C. Danto, *Narration and Knowledge* (New York, 1985). After the first version of this paper was completed, David Carr's recent book, *Time, Narrative, and History* (Bloomington, Ind., 1987), came to my attention. His approach will be discussed in the final section.

19. Fabian, *Time and the Other,* pp. 17–18, and also Index, under "Cosmology, political."

20. Consider the ancient literary genre of "wondrous" ethnographic "curiosities." What could be more predictable than these stereotypical lists, which one generation of writers copied from another? And doesn't predictability make stories of "suspense" so delightful?

21. Carr, *Time, Narrative, and History,* p. 182.

22. Ibid., pp. 182–83.

23. Among others he has in mind Louis O. Mink, "Narrative Form as a Cognitive Instrument," in *The Writing of History: Literary Form and Historical Understanding,* ed. R. H. Canary and H. Kozicki (Madison, Wis., 1978), and Hayden White's work, especially "The Value of Narrativity." The review of the literature on narrativity in Carr's Introduction is a very valuable part of his study.

24. In a way he "ontologizes" narrative by making it the structure of being human. I am not sure whether it has occurred to him that his position comes very close to all those conceptions—expressed in myths of creation—according to which reality, being, is grounded in a primordial "story."

25. Giving a reason that shows that he does have a notion of the political implications of narrative, Carr says that Hegel's failure "is not a theoretical but a practical one; it [Hegel's philosophy of history] should be understood not as a putative science but as a kind of world-political rhetoric, and its problem is not that it makes false predictions or implausible claims about the end of history, but that it is not able to constitute a community of humanity by telling a persuasive story about it" (Carr, *Time, Narrative, and History,* p. 159).

26. Ibid., p. 184.

27. Ibid.

COHEN: *La Fontaine and Wamimbi*

I am grateful to George Martin for his assistance in working with the arguments in this essay. At an earlier stage, Jonathan Lewis gave valuable assistance to the specific comparison of the La Fontaine and Wamimbi texts. Stephen Bunker, Fred Cooper, Milad Doueihi, Gillian Feeley-Harnik, Ashraf Ghani, Suzette Heald, Gabrielle Spiegel, Michael Twaddle, and Katherine Verdery have been helpful in innumerable ways in the development of this work. Members of The Seminar in the Department of History, Johns Hopkins, gave a first draft a thorough and helpful going over. Members of the Africa Workshop at the University of Chicago also provided many comments of value.

1. Johannes Fabian, *Time and the Other: How Anthropology Makes Its Object* (New York, 1983), p. 41. Perhaps Fabian does not mean for us to read this argument as it seems to read, for he provides sufficient evidence of discourses on the logic of Time in anthropological study to build a case that he is wrong. More importantly, we are, with this apparent contradiction, exposed to the question of how anthropology gained this attribution. At particular moments, and in some branches, programmatic declarations were made about the distinctions in purpose between historians and anthropologists. Indeed, in the process of formally excluding the study of the past from their work, anthropologists may have become quite sensitive to Time as concept, construct, and problem, whereas historians, claiming the past as their object, may have accepted the given, ongoing, unproblematized sense-of-Time. Ashraf Ghani has attempted to make me understand this point.

2. Fabian means a deeply etched "cultural, ideological bias toward vision as the 'noblest sense,'" such that "to 'visualize' a culture or society almost becomes synonymous with understanding it" (p. 106).

3. Most notable are the papers collected together in James Clifford and George E. Marcus, eds., *Writing Culture: The Poetics and Politics of Ethnography* (Berkeley, Calif., 1986).

4. I am grateful to George Martin for this reading.

5. Clifford reads this same photograph in a somewhat different way: "Our frontispiece shows Stephen Tyler, one of this volume's contributors, at work in India in 1963. The ethnographer is absorbed in writing—taking dictation? fleshing out an interpretation? recording an important observation? dashing off a poem? Hunched over in the heat, he has draped a wet cloth over his glasses. His expression is obscured. An interlocutor looks over his shoulder—with boredom? patience? amusement? In this image the ethnographer hovers at the edge of the frame—faceless, almost extraterrestrial, a hand that writes" (Introduction, *Writing Culture,* p. 1). One might note that it is likely that the informant/interlocutor was present in this photograph, and in proximity to the field-worker, by arrangement of the field-worker rather than the other way around.

6. James Clifford, "On Ethnographic Allegory," in Clifford and Marcus, eds., *Writing Culture,* p. 117. There is, nevertheless, the question "*Who* has the power to inscribe?"

7. Ibid.

8. The debates in anthropology over what texts constitute authentic anthropology or ethnography form a curious and illuminating literature, for they typi-

cally develop over works that present themselves as actually, or practically, "native" in their origins. A work that represents itself as an authentic native account of a culture loses its claims to plausibility as anthropology. In one sense—and to violate the sensitive elements in these debates—to succeed as "participant observer" through transcending the limits of experiencing another culture may be to fail as an ethnographer. For an interesting discussion of a case in which anthropologists have debated whether published works constitute true ethnography, see the paper by Mary Louise Pratt in the Clifford and Marcus collection, pp. 28–32.

9. And they miss the self-view and irony in that their project requires them to be impresarios to this program of subverting the conventional ethnographic mode.

10. See David William Cohen, "The Production of History," position paper for the Fifth Roundtable in Anthropology and History, Paris, July 1986.

11. Jean La Fontaine, *The Gisu of Uganda* (London, 1959).

12. Paradoxically, while La Fontaine's principal attentions among the Gisu were focused on initiation ritual, her mentor Fortes found no initiation rituals among the people he studied, the Tallensi.

13. It is perhaps appropriate to offer some descriptive words about the Gisu (also "Bagisu" [pl.] and "Mugisu" [sing.]), whose homeland is Bugisu in eastern Uganda. Over several years of writing from her 1953–55 field research, Jean La Fontaine developed a brief, routine statement—which she evidently conceived of as sufficiently introductory—on the people and area she studied: "The Gisu are a Bantu-speaking people who live on the western slopes of Mount Elgon, an extinct volcano which lies across the Kenya-Uganda border, some fifty miles north of Lake Victoria. They number about a quarter of a million and form roughly two-thirds of the population of Bugisu District in the Eastern Province of Uganda" (*Gisu,* p. 9). A similar statement appears in Jean La Fontaine, "Witchcraft in Bugisu," in *Witchcraft and Sorcery in East Africa,* ed. John Middleton and E. H. Winter (London 1963), pp. 187–88: "Since the Gisu have been described in detail elsewhere [*The Gisu of Uganda*], it is only necessary to give a brief recapitulation of the ethnographic background to the discussion. This Bantu tribe, numbering about a quarter of a million, practices settled agriculture on the fertile soil of Mount Elgon, an extinct volcano on the eastern border of Uganda. They are densely settled and land shortage is an old problem that is now becoming acute. The framework of Gisu social organization is a segmentary localized lineage system of which the basic political unit is the village, settled by members of a minor lineage. The minor lineage is segmented into minimal lineages which are the exogamous, property-controlling groups. Villages are not compact settlements but a defined tract of territory over which the homesteads of the inhabitants are scattered. Each village is roughly divided into neighbourhoods." Cf. John Roscoe, *The Northern Bantu* (Cambridge, Eng., 1915), p. 161: "*The Bagesu one of the most primitive of Bantu tribes.* The Bagesu are a Bantu tribe living upon the eastern and southern slopes of Mount Elgon. They are a numerous people when judged by the numerical standard of other African tribes, being estimated at not less than a million souls. They are a very primitive race and stand low in the human scale, though it is somewhat difficult to understand why they should be so intellectually inferior, surrounded as they are by other Bantu tribes much more

cultivated and civilised than themselves." Also cf. John Roscoe, *The Bagesu* (Cambridge, Eng., 1924), p. 1: "The Bagesu tribe on Mount Elgon is one of the most primitive of the negro tribes of Africa, and was driven from the plains to the east of Mount Elgon by the attacks of the Masai and Nandi. To escape the ravages of these warlike tribes, the Bagesu fled to the mountain, only to find that on the lower slopes they were subject to periodical raids by the Abyssinians and those tribes who inhabited the borders of Abyssinia. They therefore made their way to the less easily accessible heights." Also cf. Suzette Heald, "The Ritual Use of Violence: Circumcision Among the Gisu of Uganda" (in press): "The Gisu, numbering some 500,000, are Bantu-speaking agriculturalists living on slopes of Mount Elgon on the Ugandan side of the border with Kenya." Finally, cf. Stephen G. Bunker, *Peasants Against the State: The Politics of Market Control in Bugisu, Uganda, 1900–1983* (Urbana and Chicago, 1987), p. 32: "The Bagisu, a Bantu people numbering over 500,000, occupy Bugisu District, which covers 1,170 square miles of eastern Uganda and includes the western and southern slopes of Mount Elgon. The earliest Bagisu settlements were on the mountain, but during the last century many of the Bagisu moved down onto the plains. They first located in the southwest and then moved around the mountain to settle the central and later the northern areas."

14. Jean La Fontaine, *Initiation: Ritual Drama and Secret Knowledge Across the World* (New York, 1985).

15. Clifford, "On Ethnographic Allegory," p. 114.

16. For a valuable and suggestive discussion of the distinctive features among variant past-tenses, see Emile Benveniste, "The Correlations of Tense in the French Verb," in his *Problems in General Linguistics,* trans. Mary Elizabeth Meek (Coral Gables, Fla., 1971), pp. 105–15, and particularly his argument concerning the different qualities of aorist and preterite tenses in French, pp. 207–15 and notes, pp. 308–9.

17. This work was apparently circulated only in a mimeograph edition.

18. As indicated, throughout this comparative reading the La Fontaine text appears on the left, the Wamimbi text on the right. It should be noted that the La Fontaine segments—with the exception of a few sentences omitted—are continuous.

19. I am grateful to Professor Stephen Bunker for these observations, which come from his fieldwork in Bugisu, done at approximately the time that Wamimbi was drafting his manuscript.

20. I am particularly grateful to Gabrielle Spiegel for bringing attention to the very different handling of emotion in the two sources, which she observed in reviewing an expanded array of texts.

21. An equally elaborate rendering is to be found in the unpublished "History of Circumcision in Bamasaba Tribe," by A. K. Mayegu. In her thesis, La Fontaine presents a list of "age-set names" from 1817 to 1952; her list segments into two and then three district lists in 1861 and 1917, respectively. See Jean La Fontaine, "The Social Organization of the Gisu in Uganda with Special Reference to Their Initiation" (Ph.D. diss., Cambridge University, 1957), pp. 295–96. An example of Wamimbi's annotations of his age-set list is provided below:

Year:	*(Age-set)*	*Name:*	*Explanation:*
1956		Cohen Mutesa Namboozo	Sir Andrew Cohen the Governor of Uganda was the first Governor of Uganda to witness circumcision ceremonies. Some people call it *Mutesa* because the Kabaka Mutesa II visited Bugisu during this year after his first deportation. Some people call this period Namboozo a nick-name for La Fontaine, a Sociologist who stayed in Bugisu for four years doing research on *The Gisu of Uganda*.

22. David Henige, *The Chronology of Oral Tradition* (Oxford; 1972), p. 95.

23. Ibid., p. 96. The proverb is quoted by Henige from S. G. Champion, *Racial Proverbs* (New York, 1931), p. 365.

24. Henige, *Chronology,* p. 103.

25. David Henige, *Oral Historiography* (New York, 1982).

26. The discussion here is stimulated by Dominick LaCapra, "Rethinking Intellectual History and Reading Texts," in *Modern European Intellectual History: Reappraisals and New Perspectives,* ed. Dominick LaCapra and Steven L. Kaplan (Ithaca, N.Y., 1982), pp. 47–85.

27. Carlo Ginzburg, *The Cheese and the Worms: The Cosmos of a Sixteenth Century Miller* (Baltimore, Md., 1980), pp. 58–59.

28. Roger Chartier, "Intellectual History or Sociocultural History? The French Trajectories," in *Modern European Intellectual History,* ed. LaCapra and Kaplan, p. 30.

29. The discussion that follows is based on Stephen Bunker's field notes and copies of Wamimbi's memoranda and correspondence. These were passed along to me by Stephen Bunker, for which I am very grateful.

30. Eleven months later, in a memorandum prepared for Bunker, Wamimbi wrote that the Masaaba Historical Research Association "was founded on 6th February 1954 in St. Peter's College, Tororo in Eastern Uganda by Mr. George Masaaba Wamimbi, who has been its President-General ever since." One should note here the work of Michael Twaddle, who met George Wamimbi during his research in Mbale, Uganda, in the 1960s. Twaddle found that "during the 1950s a number of young Gisu came together to form the Masaba Research Society in order to collect and collate local traditions of origin relating to Mbale and its ethnic environs. But increasingly this society has become a collection of straightforward antiquaries as its political intentions have been frustrated by official delays. Today its members are attempting to standardize the Lumasaba language, record old circumcision songs, and collect materials for an ambitious *History of Bugisu* to be edited by the Society's President, George Wamimbi." Michael Twaddle, "Tribalism in Eastern Uganda," in *Tradition and Transition in East Africa,* ed. P. H. Gulliver (London, 1969), p. 202.

31. One notes an effort among some Bagisu to replace such terms as "Bagisu" and "Lugisu" with "Bamasaba" and "Lumasaba." This was perhaps viewed at the

time as needed and appropriate modernization and as a tactical separation of the new literature from the older studies such as Roscoe's, noted above.

32. This recalls the passage cited above, where Wamimbi states: "At this moment the onlookers keep up shouting and yelling and the atmosphere is mixed with great excitement full of joy, worry and great expectations. The noise made by the spectators is extremely deafening." Here and there in *Modern Mood,* Wamimbi resets such images against the surveillance and regulation of the state, and also, in a sense, against the cold descriptive form of La Fontaine's ethnographic text. Wamimbi's evocations of boisterous, carnivalesque opposition to authority tempt one to consider Mikhail Bakhtin's writings on the expressions of popular culture and resistance. See his *Rabelais and His World,* trans. Hélène Iswolsky (Cambridge, Mass., 1968).

33. I am most grateful to Katherine Verdery for drawing my attention to this question: to paraphrase Stanley Fish, "Is there a native in this text?"

34. This type of approach is somewhat developed in La Fontaine's dissertation, but more fully evolved in Victor Turner, "Symbolization and Patterning in the Circumcision Rites of Two Bantu-Speaking Societies," in *Man in Africa,* ed. Mary Douglas and Phyllis Kaberry (London, 1969). Turner did research on circumcision in southern Bugisu.

Index

Library of Congress Cataloging-in-Publication Data

Chronotypes : the construction of time / edited by John Bender and David E. Wellbery.
p. cm.
Includes bibliographical references and index.
ISBN 0-8047-1910-1 (cloth : acid-free paper)
ISBN 0-8047-1912-8 (paper : acid-free paper)
1. Time. I. Bender, John B. II. Wellbery, David E.
BD638.C48 1991
115—dc20 90-26967
CIP